RENEWALS 691-4574

DATE DUE

333			
438			
525			
507			
4268			
7:59			
3:59			
11:06			

Demco, Inc. 38-293

BASIC DILEMMAS IN THE

SOCIAL SCIENCES

Hubert M. BLALOCK, Jr.

BASIC DILEMMAS IN THE SOCIAL SCIENCES

SOCIAL SCIENCES

SAGE PUBLICATIONS Beverly Hills London New Delhi

For information address:

SAGE Publications, Inc.
275 South Beverly Drive
Beverly Hills, California 90212

SAGE Publications India Pvt. Ltd.  SAGE Publications Ltd
C-236 Defence Colony 28 Banner Street
New Delhi 110 024, India London EC1Y 8QE, England

Printed in the United States of America

Library of Congress Cataloging in Publication Data

Blalock, Hubert M.
    Basic dilemmas in the social sciences.

    1. Social sciences. I. Title.
H61.B478   1984   300   83-24590
ISBN 0-8039-2021-0

FIRST PRINTING

CONTENTS

Preface

Over a period of at least 15 years I have become increasingly disturbed by a number of patterns and trends that characterize my own discipline of sociology. I have also become aware of similar patterns in political science and at least some fields of other social science disciplines as well. In the most general terms these patterns suggest that if we ever possessed a sense of direction and the makings of an intellectual core in each of our disciplines, we appear to be moving increasingly in a haphazard fashion with little or no consensus—or even much concern—about the kind of intellectual products our aggregated actions are bringing about.

My experiences within the American Sociological Association have done nothing to soothe this increasing sense of concern about our highly disjointed efforts. Indeed, as Chair of the Program Committee, I found that trying to organize a set of some 200 individual sessions into a reasonably coherent overall program became a hopeless objective. Responding to numerous requests for all kinds of sessions in proposed new subfields or involving highly concrete topics of immediate interest to a dozen or so colleagues became an equally disturbing experience. It also became clear that many sessions and sub-subfields within sociology have been established more for expedient reasons that for academic ones. We evolved sessions devoted to the black community, and then we focused others on Chicanos. But then the Puerto Rican members also demanded a session, and still others were requested so as to deal with each of our other minorities, including not only all of our separate Asian-American ethnic groups but Eastern European ones as well. Finally, I received a request for a session on black women. What kind of "balkanization" of the field of race and ethnic relations would this eventually produce and with what long-term implications for sociology? Fortunately, some of the worst excesses of these political processes are now behind us. But there has been a cost, as well as some wholesome effects. The question is whether some of these trends can be reversed without giving up the beneficial democratizing patterns that accompanied them.

The immediate impulse that resulted in my decision to put my concerns into book form came, however, as I attempted to develop a very general lecture intended primarily for a lay audience, many of whom would be unsympathetic to and uninterested in our internal disputes. The tack I decided on was to stress that social scientists face a number of fundamental complications that are not of our own making. These complications, when occurring simultaneously, will almost inevitably result in confusion, ambiguities, and legitimate disagreements as to intellectual priorities and optimal strategies. These fundamental problems that are in some sense "out there" then interact with our own characteristics, as scholars and human beings, to create a series of dilemmas and hurdles that can only gradually be understood and overcome. Our lack of financial resources and the scattered and short-term nature of most of our research efforts only compound our difficulties, as does the confused division of labor between and within our social science disciplines.

Although written very nontechnically, this book is intended to serve as a serious effort to assess what is wrong with the social sciences and what steps may be taken to improve the situation. As implied, the thesis is that our problems are only partly of our own making and that we can hardly expect to succeed unless we can also achieve a much more satisfactory understanding of the implications of the simultaneous occurrence of a number of complicating "facts of life" that will remain to confront us, whether we recognize them explicitly or not. Thus there are a series of dilemmas we must face that are primarily methodological in nature, or that stem from a combination of methodological difficulties and our own limited resources in terms of time, money, and organization. This suggests that even under relatively ideal conditions one must expect a number of genuine disagreements over strategies, priorities, and philosophical orientations.

There are other problems as well, however. Many of these concern our more human difficulties as researchers who inevitably must cut corners, as faculty or members of research teams who seek promotions and recognition, or as citizens or scholars with ideological axes to grind. Some of these problems will be hinted at in the introductory chapter and the four succeeding chapters that deal primarily with methodological issues. But my principal discussions of some of our human and organizational problems will be reserved for the final two chapters. The essential argument will be that unless social scientists learn to pull together more effectively, while also facing up to the methodological implications of simultaneous complicating factors, we fully expect to witness a continuing fragmentation of our disciplines. Indeed, we may expect to create a brainless monster whose development becomes totally beyond our control.

Perhaps the situation is not as hopeless as the monster analogy suggests. In any event, however, I believe it will do no harm to make a strenuous effort to examine those common methodological problems that we all confront, to strive for a much better consensus on concepts and definitions, and to examine carefully the nature of the scholarly norms we have evolved. We need to discourage those aspects of our collective behaviors that lead to irresolvable disputes, and the simplistic discounting of arguments made by "rival" schools, or that are concerned more with a need to sell ourselves to lay audiences than to communicate effectively with one another. In short, we must recognize that although many of the complications and hurdles we face are real, our own shortcomings and diverse backgrounds and orientations may be further compounding our problems.

This short book represents one more effort, then, to confront the perennial problems that all social scientists face: Why aren't we doing better, and how can we improve? The ideas contained in the book are certainly not new. My hope, however, is that they have been packaged in such a way that readers will agree with my most general conclusions that, although the methodological difficulties we face are indeed complex and place important limitations on what we may hope to achieve in the short run, a soul-searching effort to assess their implications and then act on them in a reasonably concerted fashion not only represents our best hope. It also holds promise of gradually achieving the kind of cumulative knowledge that many of us believe can be genuinely useful in improving the lot of mankind.

CHAPTER
1

The Challenge

It does not require a fine-tuned measuring instrument to show that many social scientists are not happy with the way things are going. We see declining student enrollments and an economic retrenchment that threatens many of our members with unemployment or reduced opportunities for productive research and teaching. We want to have an impact on policymaking. Yet the opportunities have not been forthcoming except on a very limited and selective basis. Why won't anyone listen to us? And what is happening to our individual disciplines? Are we becoming a polyglot category of scholars held together primarily by the negative fact that most of us have been clumped together into specific social science departments?

What are the intellectual cores of the several social sciences? How can we guide their development and prevent the fragmentation of each of our disciplines? Will the almost inevitable tensions produced by a slow economy and decreasing job opportunities become closely linked with intellectual and ideological cleavages, exacerbating scholarly debates as to how best to proceed? Will this effect the quality of our teaching as well as the coherence of our course offerings and the ways we present ourselves to students? These are serious times and thus it becomes especially important to take stock. It may turn out that there is indeed no great cause for concern, but if so, no harm will come of it.

The thesis of the present work is that many of our problems are created by two types of ingredients. On the one hand are a number of very tough intellectual challenges stemming from the complexity of the social world we are attempting to study. On the other there are our own diverse intellectual backgrounds, interests, and behaviors. If we are to understand our present difficulties, we must examine the interplay between these two kinds of factors: ourselves and the limitations imposed on us by the nature of our subject matter. Who are we? What are our expectations and intellectual orientations? What do we want to explain and why? How do we organize ourselves to do so? Just what are the features of social reality that create problems for us in terms of complexities and ambiguities that are difficult to resolve? What are the kinds of social issues that seem to require immediate answers and that produce tensions between the desire to understand phenomena as thoroughly as we can and the practical necessity of acting on the basis of ambiguous empirical evidence?

We have been attracted to our disciplines for a variety of reasons, but it is essential in this connection to realize that most of us are something of a cross between scientists and humanists. Many of those who take a "hard science" orientation were initially drawn to the social sciences because they were deeply disturbed about some aspect of society they wanted to see corrected. I selected sociology because I was concerned about the treatment of black and other minorities in American society. Some persons are attracted to demography because they believe that worldwide population problems must be solved if human beings are ever to achieve a reasonably satisfactory life on this planet. Others are concerned about the stability of families or the prevention of delinquency.

Other social scientists have a much more hostile orientation toward the sciences or else believe that the primary role of social scientists is to act as social critics. We should serve as analysts whose principal task is to appraise existing social institutions: to point out their functions, shortcomings, and sources of support, with a view to modifying them in one or another direction. Or we should seek an understanding of human behaviors that is more akin to that of the poet or artist rather than that of the scientist. It is argued that science merely serves the interests of those who sponsor the research and thus is fundamentally a conservative force that should be attacked rather than emulated. Not surprisingly, there have been numerous debates concerning the legitimacy of the social science enterprise itself. For instance, the 1960s was a traumatic decade for several of the social sciences—particularly sociology, cultural anthropology, and political science—because the credibility of the sciences was under attack by a substantial proportion of our members.

Thus social scientists are far from homogeneous in terms of intellectual orientations. Indeed, we are just about as heterogeneous a grouping of scholars as one would ever expect to encounter. In some instances differences within a given discipline are far greater than those between fields, making it easier, for example, for many sociologists and political scientists to communicate with one another than to do so with their own colleagues. This extreme heterogeneity has some positive consequences but also a number of not-so-happy outcomes. Put together an extremely heterogeneous group of scholars and an elusive social reality we are attempting to study, and you have the ingredients of a very confusing intellectual atmosphere.

A COMBINATION OF COMPLEXITIES

It may be relatively easy to deal with complexities as long as they occur one at a time. Generally, complexities produce situations in which there are too many unknowns relative to the availabile pieces of empirical information. By making simplifying assumptions of one kind or another one may reduce this set of unknowns to manageable proportions. If their numbers are not too large, it may also be possible to collect additional data to provide consistency checks, so that the reasonableness of the assumptions may be assessed indirectly. When two or more complexities are introduced *simultaneously*, however, the situation may get out of hand and become intractable without some rather strong a priori assumptions that may have the effect of doing away with the problem altogether. For instance, when one allows for unknown lag periods and measurement errors in dynamic analyses, there may be too many unknowns unless one assumes, a priori, that either the measurement errors are absent or that the lag periods are known.

Perhaps the most fundamental reason why we experience so many ambiguities in interpreting the results of social research and in testing theories with real data is that we are usually confronted by such simultaneous complications. This is not the place to become involved with the technical issues, but at the outset it is advisable to mention briefly some of the most pervasive complications that create the kinds of formidable problems that will concern us in later chapters. The major point is that the ambiguities that are created by these simultaneous complications have important implications in terms of the debates that emerge, the ways in which we organize to collect our data, the general research and theory-building strategies we endorse, and, of course, the rate at which our knowledge base accumulates or fails to accumulate. Our morale, expectations, and professional norms are also affected indirectly. It cannot be overemphasized, then, that these "real-world" complexities, if taken as givens, have a truly major impact on the entire social science enterprise. What are some of them?

First, virtually all social processes are far more complex than we often realize. Our explanatory laws must therefore be both multivariate and probabilistic; furthermore, the variables we usually select as "independent" are, themselves, often highly intercorrelated and influenced by the variables we are attempting to explain. It appears as though the simultaneous equations that would be needed to account for most reasonably complex social phenomena will need to contain upwards of 20 variables and, realistically, as many as 50 to 100. And even were we able to specify models of this degree of complexity, there would still be error terms needed to account for the remaining indeterminacy in the system.

Second, measurement problems in the social sciences are formidable. Physical scientists are, of course, used to the indirect measurement of postulated properties such as mass, heat energy, or electrical charge. But the assumptions needed to link operational measures to these physical properties are much more precise and justifiable than is possible, say, when a social scientist wishes to infer the "power" of a corporate executive on the basis of the responses of other parties. A very complex auxiliary measurement theory is often needed in the social sciences to understand this linkage problem. Our present knowledge as to how to construct and analyze data with such measurement models is developing, but it is not yet sufficient to provide really definitive guidelines for measurement. In effect, measurement problems produce a host of unknowns that must be added to those involved in one's substantive theory. Sometimes the number of such unknowns, relative to the amount of empirical information, creates a hopeless situation that cannot be resolved without a combination of additional information and untested a priori assumptions.

Third, as is also common within the physical sciences, rates of change in social phenomena are sometimes far too rapid to be studied with present resources, sometimes they are far too slow. Many important data are simply lost in history and can never be obtained at any cost. Ideally we might like to have detailed information about individuals over their lifetimes, but ethical and practical obstacles to the collection of such data are overwhelming. Furthermore, many social phenomena are changing all at once and continuously, making it difficult to infer temporal sequences or to pin down dates of their onset or termination. True, there are discrete "events," but often these readily identifiable phenomena are of interest only as indicators of the variables of greatest interest to us. When there is multiple causation, feedbacks of varying and unknown durations, and continuous changes in a large proportion of one's variables, and when measurement errors of unknown magnitudes also exist, one can begin to imagine the magnitude of the difficulties one faces in attempting to disentangle causes and effects.

Fourth, there is a tremendous variety of behaviors and other phenomena we wish to explain. Are there any systematic ways in which we can reduce the complexity of the situation by identifying a much smaller number of more inclusive variables or social processes, taking concrete events as special instances? Certainly we may come up with rather general labels for these variables and processes, but problems of measurement comparability loom large when we do so. How, for instance, can one compare aggressive behaviors, political participation, or worker alienation across a variety of contexts? Must our measurement operations change if we move from a very simple laboratory situation to a more complex one—and if so, then how do we distinguish measurement noncomparability from substantive differences across these settings? What strategies can we use to define manifestly different behaviors, all of which are presumed to have similar consequences, when these very consequences are influenced by other factors besides the behaviors we are attempting to measure? It is perhaps a truism that all analyses require simplifications and abstractions, but of what kinds and according to what rationales? If we do not attempt such simplifications, what are we left with, apart from descriptive materials? *Someone* must eventually try to add up such descriptions to provide more general principles. But it is far from obvious how to proceed.

Fifth, the reality we must deal with is often fuzzy or imprecise. Even where we wish to obtain precise measures we may therefore be unable to do so without making arbitrary decisions, each of which may yield somewhat different conclusions. For instance, the boundaries of an urban area or informal clique may be indeterminate. If we select arbitrary boundaries—say those coinciding with political lines—we may obtain a precise measurement of residential segregation, but different boundaries would undoubtedly yield somewhat different results. There is therefore a kind of social science analogue of Heisenberg's Uncertainty Principle in operation, placing limits on the degree of precision of our measuring instruments and therefore the accuracy and adequacy of our scientific predictions. If we simply make measurement decision by fiat, asserting that we must use only the available data, problems of measurement comparability will loom very large.

Finally, there is no obvious way to divide up the labor, either among the separate social science disciplines or within any one of them. To be sure, there are some arbitrary disciplinary boundaries that have been arrived at through historical accident. The problem is that multiple explanatory variables spill across the currently defined domains of each of the social sciences. One cannot arbitrarily omit any large fraction of these variables without doing injustice to social reality. If one social scientist uses a set of, say, ten explanatory variables and a second investigator a partially overlapping set of six, one can rarely fit the two studies together because of the missing information in each study. If each investigator also has used somewhat different measures on different populations at different points in time, the number of implicit or explicit assumptions needed to lace the studies together begins to mount exponentially. Imagine, then, the difficulty in adding up the findings of disparate studies, each with a somewhat different focus and employing different methodologies in widely differing contexts. Given the rather high intercorrelations that often exist among potential explanatory variables, the setup is wide open for ideological and disciplinary biases to influence the interpretation.

The situation is disturbing enough, but we must add another set of ingredients. Most social research—particularly *data collection*—is very expensive and time consuming. Often our data are collected by someone else, such as a government agency. Sometimes we are fortunate enough to have roughly comparable time-series data over a rather long period, perhaps several decades. But the more usual situation involves single investigators or very small research teams collecting their own data with modest and short-term funding. Nor is this funding equally available in all fields or for all purposes, with the result that gaps in our knowledge are distributed unevenly. Let me note a few specific difficulties in this respect.

From many standpoints experimental designs represent an ideal way to proceed. Social scientists can only experiment on certain kinds of individuals under restricted conditions, however. Experiments are typically conducted on powerless people—schoolchildren, introductory psychology students, or hospital patients. We do not experiment on the Washington State Legislature or the board of directors of Boeing, as much as we might enjoy doing so if this were feasible. Nor can experiments be carried out over more than a few days or involve risky or noxious stimuli. Therefore, experimental manipulations usually involve very minor and short-run changes that are difficult to measure or assess over some prolonged period. Nor are there any scientifically sound bases for generalizing from such powerless or cooperative subjects to bank executives, professional criminals, or political elites. The number of true experiments that can be conducted is thus very small, constituting a narrow band among the topics in which social scientists are interested.

Social surveys are extremely expensive and beyond the means of virtually all individual investigators. They must therefore be financed by someone, and of course these outside parties may "call the shots" in terms of what is and is not investigated. One of the outcries on the part of radical and minority social scientists during the 1960s was that elites are seldom studied in this way. More often a so-called problem is defined to exist in terms of some undesired *minority* behavior, such as poor school performance, above-average illegitimacy levels, or high crime and delinquency rates. The problem is thus defined to be one of changing the *minority* rather than elites or social institutions. These objections of radical and minority social scientists were sometimes overstated, in an obvious attack on science-oriented quantitative research. But there was considerable truth to them, and the arguments remain valid today. If research is expensive, we can fully expect that financing will be selective and that many important questions will not be studied. Some subareas will be financially starved whereas others will be overfed, and this will have little or nothing to do with their social importance or their relevance for the advancement of social science knowledge.

Practically all funding in the social sciences is short-term in nature, so that replications or prolonged time-series analyses are only rarely accomplished except in a few well-financed, popular areas where an obvious payoff may be expected. If one adds to this the fact that most social research is conducted by individual investigators who must make tenure at their universities or apply each year for new grants, and who can hardly afford to collect data over a 20-year period, it becomes obvious that many kinds of important analyses simply will not be made. It is not as though one could carefully collect data over a long period but with a relatively high degree of assurance of a later payoff. Most social research is risky and unlikely to yield dramatic results; therefore a premium is placed on relatively quick projects. Someone who invests five years or so on a major survey usually will have learned a lesson—namely, not to do it again without a considerable recuperation period.

Other kinds of data-collection methods are far more exploratory in nature and much more suitable for small-scale, individual efforts. But here there is a tendency to select research topics for idiosyncratic reasons and to use research strategies that are difficult to replicate. The result is usually a series of disjointed efforts—some of which may be highly insightful—that do not flow naturally into larger-scale, more systematic research despite general agreement that this would be desirable. One reason is that social scientists are not organized to do team research, except on an ad hoc or short-term basis.

THE ANTI-SCIENCE ATTACK

The social sciences, lodged as they are between the natural sciences and humanities, have almost inevitably become a battleground over the suitability of natural science models and approaches to the study of human behaviors and social processes. The intensity of the debates waxes and wanes, sometimes owing to changes in our environment—as for example, the Vietnam and Civil Rights periods of the late 1960's and early 1970's—and sometimes simply because each new generation of social scientists raises the issues anew and then becomes tired of the ensuing arguments. Yet the issues are admittedly important ones. On some occasions positions are stated so dogmatically that motives are transparent, as for example the effort to discredit the use of any data that might effectively challenge a particular viewpoint. But there also may be occasions when the debate becomes more productive by bringing to light issues that have been neglected or set aside because earlier answers had proved ineffective.

Those of us who believe in a science-oriented approach, suitably modified to the special problems faced by the social sciences, must continually examine some of the most telling points made by humanists and others who remain skeptical of natural science approaches. What have been the principal arguments on both sides? Which of the issues appear resolvable through a process of improving our understanding of the logic of the scientific method and its limitations? In the discussion that follows I shall attempt to state many of the most common objections to a science-oriented viewpoint, without attempting to credit specific authors with having been the first to raise each issue and without attempting to locate the most eloquent statement of each of the positions taken.[1]

There is a whole series of arguments that boil down to the thesis that human behavior is a fundamentally different kind of phenomenon, one that cannot be studied by scientific means or by any approach that attempts to break a totality down into parts that can by analyzed separately. "The whole is greater than the sum of its parts." Human behavior must be seen in its totality and must be experienced firsthand to be understood. If we attempt an analysis, which almost by definition requires some effort to decompose a phenomenon, we miss the true essence of human life, which can only be known in some other way. A scientific orientation, at least as applied to the study of human behaviors, stifles creativity, verstehen, or some other means of comprehending the full meaning of the phenomenon. This type of challenge has considerable appeal, especially to the poets and artists among us. To others, it smacks of unadulterated mysticism, with a pretense that there is some sort of deeper form of understanding that can never be communicated to others. We all sense what the proponent of this viewpoint is getting at, since scientific understanding must always be imperfect, since only a limited number of questions can be answered by scientific means, since we must ultimately rely on our own senses as our most direct form of information about the outside world, and since questions of deeper meaning are much easier to ask than to answer. What is at issue here may rather simply be one of our aspiration levels. If so, it becomes essential to understand the limitations of the scientific method and to recognize that it is only at relatively advanced stages of a science that rapid progress and genuine knowledge accumulation become obvious. Those who want to obtain their knowledge quickly and who desire a high degree of certainty and irrefutability to their arguments will be doomed to disappointment with a scientifically oriented approach. The situation is most frustrating when one is dealing with a highly important social problem for which immediate answers are desirable and yet are not forthcoming.

There is, however, a far less idealistic motivation behind any anti-science claims of this nature. For instance, if one must experience a phenomenon first-hand before understanding it, this would seem to imply a rather rigid division of labor. Only a member of a minority can understand the impact of institutional racism. Therefore, whites have no business studying minorities or even the field of race relations. Pushed to the extreme, such a position implies that no one can study anything other than his or her unique experiences, and no one is qualified to "add up" or make comparisons across a number of such unique phenomena. In effect, one is told to keep out of another's territory because he or she lacks the necessary insight to study the phenomenon in question. The underlying vested interests in making such a claim cannot be ignored, especially when a relatively new or young group of scholars is challenging the "establishment."

Yet these arguments cannot be dismissed totally. Since no one has been able to pin down just how insights are obtained, firsthand experiences are obviously important. So-called marginal persons, who have experienced conflicting demands or been placed in contrasting situations, are in an excellent position to analyze phenomena that others take for granted. Those who have experienced discrimination or any other type of abuse are in a much better position to understand the reactions of others who have been similarly exposed. As researchers they would have special entree to persons like themselves, although they are also likely to share the biases of these actors. The obvious implication is the need for a division of labor and the sharing of insights and critiques with a view to correcting misleading interpretations, noting omissions and biases, and suggesting specific programs of research that will enable us to fill in the gaps and correct existing biases.

Other critiques of science-oriented work in the social sciences point to the extremely complex nature of social reality, the uniqueness of specific historical events, or free will and indeterminism as constituting such formidable roadblocks that it is foolhardy to attempt scientific analyses or generalizations of more than a limited nature. The existence of historical facts may be acknowledged and the study of such facts taken to be a legitimate intellectual endeavor for its own sake. But it is believed that theoretical propositions purporting to explain such facts cannot be stated for a number of reasons. Or such propositions, if stated, will be of limited utility because situations will be noncomparable and measurement next to impossible. Furthermore, there will be too many unknowns and uncertainties, making qualifications of such propositions totally unwieldy and in need of perpetual modification. It is argued that certain kinds of variables—especially postulated mental states and complex cultural factors—are either inherently unmeasurable or measured with such inexactitude that almost any empirical efforts to do so will be suspect and nearly worthless. Such theories as we may develop should be concerned with interpretations of specific historical events and cannot be formalized or systematized in a pseudo-mathematical fashion.

There is much to be said in favor of all of these points. Social reality is indeed complex, replications are far from perfect, measurements *are* indirect and subject to biases of many kinds, and no particular study can be expected to control for all relevant variables. Human behavior, being motivated by multiple goals and enacted under diverse conditions, will be imperfectly predicted by taking only a portion of these factors into consideration. Comparability of measurements will be difficult to achieve across settings and therefore substantive findings and measurement artifacts will be difficult to disentangle. Nor will the conditions affecting particular outcomes be easy to specify theoretically, apart from indicating the time and place at which they occur. This suggests at a minimum that our level of scientific aspiration must realistically be much lower, especially when we are still grappling with understanding even the simplest of social processes that may be studied in laboratory settings.

In responding to criticisms of this type, what we need are formulations that take rather general assertions about complexities, nonmeasurability, and lack of comparability and translate them into specific complications that can be studied systematically. Here I believe that we have made considerable progress in the fields of applied multivariate analysis, measurement and scaling theory, and in the use of stochastic models that build uncertainties into our deductive theorizing about such things as decisionmaking, social mobility, and diffusion processes. We know a good deal, for instance, about the implications of high intercorrelations among independent variables, how to construct simultaneous-equation models of feedback processes, how to conceptualize measurement-error models allowing for nonrandom measurement errors, how to simplify highly complex social networks in meaningful ways, and how to evaluate the goodness of fit of such models to real data. In all of these instances the constructive step is to admit to the complications, try to specify their natures as carefully as possible, and then state a realistic set of assumptions needed to handle them.

When we follow such procedures, however, we *do* discover that simultaneous complexities may create unsurmountable obstacles in the form of too many unknowns. If this can be recognized through an analysis of the formal properties of one's theoretical model, it implies that no data whatsoever will suffice to resolve the problem without further assumptions or simplifications. If so, we know that certain kinds of disputes will be irresolvable on empirical grounds or will require the collection of data beyond those currently available. This suggests that energies might be directed more effectively elsewhere—unless, of course, one wishes to raise questions that cannot be answered or to state a theoretical position that cannot be either verified or refuted by empirical evidence. What we must admit, then, is that these complexities *do* create problems, some of which cannot be resolved with our current resources. Rather vague assertions about noncomparability, free will, or hopeless complications must be taken with a grain of salt if we are to make any progress at all. They caution us, however, not to expect too much too soon.

There also is an important sense in which such claims fail to recognize comparable complexities in the physical and biological realm, as well as the long time it has taken in most of these fields to achieve the status of a "science." In particular there are many instances in the sciences where natural forces are highly complex, making precise predictions almost impossible. Having listened to numerous accounts Mt. St. Helens's recent eruptions and noting the present uncertainties about future eruptions, I have regained a certain amount of assurance concerning our inabilities to forecast in the social sciences. Meteorology, too, is a very uncertain science for many of the same reasons—namely, the large number of parameters that would have to be provided to make accurate predictions that are also highly specific. Nor is the universe, the human organism, or the chemical composition of the earth's core a simple thing to study. More often than not, the ideal experiment exists in textbooks but is difficult to achieve in actuality. We in the social sciences are not alone in dealing with highly complex phenomena.

Finally, there have been a set of claims about the nature of science and its sponsorship that involve a number of value issues, quite apart from the feasibility of attempting a science-oriented approach. One thesis is that a push toward science predisposes one to study problems that are theoretically or practically trivial, as long as they are scientifically tractable. We examine only those variables that are easily measurable, neglecting the truly important ones. We avoid the big questions in favor of lesser ones, with the rationale that the important ones can be tackled only once the simpler ones have been resolved. Our methods dictate the problems we study rather than vice versa.

We compulsively and slavishly imitate the physical sciences, hoping that something will come of it and envying the scientist's status. We call ourselves a science merely to gain prestige, not realizing that this will never occur unless we actually achieve success with these methods. So we are sidetracked from really important concerns by what amounts to an immature effort to "be scientific." We will never achieve success unless and until we break free from this misguided effort. A social science establishment attempts to talk the science game, thereby socializing younger generations so that they will reward their elders for less than noteworthy achievements. Each successive generation thus develops a vested interest in playing the science game, since we must fool others into believing we are scientific in order to retain our status as an intellectual enterprise worthy of continued support.

Furthermore, as I have already implied, since big science requires big funding, scientists find themselves selling out to the establishment, particularly the military and giant businesses. In their efforts to get in on the funding of research, social scientists tend to accept problems as defined by such an establishment. They study "problem" populations, such as minorities and the poor, with a view to "correcting" or controlling their behaviors, rather than examining the behaviors of elites or the institutional arrangements that are predominant forces in society. Scientists tend to report to elites. If these reports imply conclusions disliked by such elites, they will either be suppressed or be safely buried in technical papers in learned journals. For the most part, and with some exceptions, science is blind to human needs, cold, impersonal, and self-serving. It "uses" relatively powerless people rather than serving them, although the practitioners and sponsors of scientific research may rationalize the endeavor in high-sounding jargon. In brief, the scientist is not to be trusted.

Social scientists who attempt to follow this path are simply naive if they believe their research will not be used to benefit the establishment groups in society, rather than the general public. Those who say they believe in "value-free" research are equally misguided. There are biases inherent in one's choice of problem, and funding patterns will reinforce these. No one can avoid such biases. Science-oriented research merely attempts to hide these biases from view.

We cannot dismiss charges such as these merely by arguing that we will attempt to state our biases—a thesis popularized in Myrdal's famous appendix to his *An American Dilemma* (1944). It *is* true that much governmental funding of research is "mission oriented" and guided by a bias toward locating "problems" in relatively powerless individuals. The evaluation of social programs is often superficial and underfunded, so that there is the definite suspicion that careful and unbiased evaluations are not really desired.

Yet the claim is exaggerated, especially when applied to funding agencies such as the National Science Foundation or National Institute of Health, both of which have encouraged a wide variety of basic research projects. In the case of the natural sciences, the thesis also neglects to mention a large number of research projects that have been funded in the name of health concerns and environmental and agricultural improvements designed to benefit large numbers of persons. Anti-science arguments were especially popular during the Vietnam War, when military research and the use of chemical defoliants dominated the news. When we recognize the huge size of the Pentagon budget for research and development, however, we see that these arguments cannot be dismissed easily. They must remain with us as a kind of perpetual note of caution.

The impossibility of a completely value-free orientation goes without saying, but it does *not* follow that the ideal cannot be approximated to varying degrees and that a careful study of specific sources of bias—as, for example, measurement errors—cannot help reduce these biases substantially. Taken as an either-or position statement, the claim that a value-free social science is impossible is often itself a smokescreen or rationale for saying that anything goes. If we cannot be truly neutral and objective in some absolute sense, then it may be claimed that we should give up the pretense. Presumably if we carry the argument to its logical conclusion it becomes permissible to use facts selectively to bolster any point of view whatsoever, neglecting all negative instances and relying primarily on the reader's ignorance of these facts or on the social scientist's ability to mislead the innocent reader.

Such a cynical view of the role of the social sciences is, of course, extreme. I would reject it outright, no matter how inadequate our existing knowledge may be and regardless of whatever hidden biases we all may have. It would obviously discredit the entire social science effort. If its implications were taken seriously, it would undoubtedly spell the end of all institutional supports for the social sciences unless they were deliberately being used in such a cynical fashion by establishment forces. A social science controlled in such a manner would indeed depart in truly major fashion from the value-free ideal.

TOWARD A CONSTRUCTIVE RESOLUTION

It is all well and good to admit problems and shortcomings and even to commit oneself to stating possible biases for readers to evaluate. But it is quite another matter to find reasonably *systematic* ways in which potential biases or misspecifications can be evaluated and even measured quantitatively to assess their seriousness. We must first admit that there can never be ironclad guarantees against unknown biases precisely because they *are* unknown. But through a combination of collective efforts by means of which one person's biases or omissions are corrected by another's insights, plus some rather rigorous guidelines for data analysis and theory construction, we can improve the situation considerably. I believe that a good starting point in this process is exposure to well-known principles of multivariate analysis and scaling and measurement theory.

When we introduce the ideas of multiple regression to students, we may also instill in them a strong sense of the necessity for intellectual integrity. When we express some dependent variable Y as a function of an arbitrary number of independent variables, we do not impose restrictions on what these variables may be or on how many to include. When we talk about how much of the total variance is explained by any one variable, controlling for the others, again we are stressing that each variable—no matter what it is—has an equal opportunity to account for this unexplained variance.

We then may go on to spell out a whole series of complications and how to handle them. For instance, we know that high intercorrelations among independent variables will create difficulties in interpretation, as well as large sampling errors. We may then alert researchers that in such instances it may be highly misleading to attempt to separate out the effects of each variable. We may also discuss the implications of measurement errors of various types, presenting techniques for assessing such errors and making corrections for them. We may also explain how nonlinearities may be examined and tested and how nonadditive joint effects may be investigated systematically rather than on an ad hoc basis. And we may stress that this can and should be done regardless of which variables have been inserted into the model.

We may also explain why assumptions about omitted variables are so crucial and what kinds of distortions are produced if these assumptions are incorrect. This then sensitizes one to the need to theorize about omitted variables or variables that have been measured poorly. With explicit equations in front of them, one's readers may then include such variables in their own studies and assess whether or not their results are sufficiently close to those previously found. Readers may also be alerted to biases produced when ordinary least-squares procedures are used with simultaneous-equation data, as well as complications produced by autocorrelated error terms or other sources of misspecifications.

Social scientists properly exposed to these and other technical issues cannot avoid coming out of the experience without some sense of concern that as many complications as feasible be examined and the results interpreted to the reader. The essential communality of all these analytic techniques, when properly employed, is that assumptions are brought into the open and stated explicitly so that they may be readily challenged. Furthermore, one is told how additional complications may be handled and variables brought into the explanatory system in an evenhanded way. One may also eliminate variables that do not work, again on objective rather than arbitrary bases.

Readers also have to be warned about pitfalls, but again this can be done without regard to the particular explanatory variables under consideration. They can be sensitized to the obvious fact that variables that are virtually constant cannot be expected to explain much variance and the implications for design noted. For example, if I select a population that is homogeneous with respect to variable X_1 but heterogeneous with respect to X_2 whereas someone else's design reverses this pattern, we may fully expect that X_2 will work relatively better in my study and X_1 in the second study. Similarly, if the random measurement error variance in X_1 (relative to the true variance) is much greater than that in X_2 (relative to its true variance), I may expect an underestimate of X_1's effects, relative to those of X_2. Readers are thus alerted to certain kinds of artifacts of measurement or research design, so that they will be sensitized to possible differences among study results.

None of this guarantees that an individual investigator's biases will not result in the selective or poor measurement of important variables. Since others will be in a position to add some variables of their own, this particular kind of problem does not appear too serious. Much more problematic, however, are disciplinary biases that result in the neglect of whole sets of factors as being outside the province of study or presumed to have negligible impacts. Here a more catholic or eclectic orientation may be encouraged so as to introduce a much wider range of explanatory variables that may have gone unnoticed even by individual investigators of differing ideological persuasions. Also, particular data collection techniques—such as survey research—may lead to the omission of certain kinds of variables, as for example early socialization experiences or contextual factors. In these instances as well readers can be sensitized to the need to use increasingly inclusive explanatory systems

Another interpretive bias may exist whenever explanatory variables are moderately intercorrelated. The kind of intellectual dispute that may then take place is often of this form: "My variables are better than yours! Yours are mere correlates or symptoms of the 'true' explanatory variables." If there are high intercorrelations among the two sets, it may be impossible to resolve such disputes empirically without much better data and larger samples. Readers may, however, be alerted to this multicollinearity problem so that they are not misled by such disputes.

Confusion may also be created by vaguely defined theoretical constructs that shift their meanings according to the circumstances. Once more, if social scientists are alerted to this type of problem and have become sensitized to the scaling and measurement error literatures, such ambiguities can often be resolved or at least brought out into the open. Constructs that appear to have a simple meaning often tend to be multidimensional in nature, a fact that may be uncovered by any number of scaling or factor analysis techniques. This is not to say that such conceptualization-measurement problems are straightforward, but at least there are relatively systematic approaches that can be used to help clarify them in a reasonably objective manner. I would argue that this is the way we must proceed if we are to answer, one by one, any reasonably specific objections to the thesis that the ideal of value-free science may be approximated within the social sciences. Complete objectivity cannot be achieved in any single study, but successive approximations can be obtained provided that assumptions are brought into the open and the rules of the game carefully analyzed and then made familiar to social science practitioners.

NOTE

1. There have been numerous discussions of the science-value issue within sociology and, I presume, in each of the other social sciences as well. A recent summary of the debate within sociology is provided by Schoombee (1983). Among the works cited in this review essay I would recommend, for a broad coverage: Becker (1971); Berger (1971); Cicourel (1964); Etzioni (1965); Lundberg (1961); Lynd (1939); Merton (1971); Mills (1959); Myrdal (1970); Weber (1949); and Znaniecki (1968).

CHAPTER

2

Complications
Produced by
Multiple Causation

The thesis of this and the following chapter is that there are a sufficient number of complex, technical, and sticky methodological problems confronting all the social sciences that we are afforded an excellent opportunity to close ranks and focus on these common concerns as matters that need to be resolved, if only gradually, if we are to achieve any reasonable set of common objectives. At the same time, however, the existence of these problems appears to be one of the reasons why we have found it so difficult to reach agreements. Almost any attempt we make to resolve them will be subject to criticism, and each resolution requires approximations and untested assumptions that many of our colleagues may criticize. Unless we recognize that the problems involved are fundamental ones creating dilemmas that can be only imperfectly resolved, these difficulties become convenient pegs on which unconstructive criticisms can be hung, rather than clearly recognized obstacles to be overcome through cooperative effort.

This is not the place for even moderately technical discussions of the problems that will be discussed briefly and of course readily available are a number of far more detailed discussions of each of them. Instead, my aim is to present a sufficient array of methodological problems that cross-cut most of the social sciences that the overall message will have been conveyed. It is that many of the difficulties we face are not of our own making. They stem from a combination of things, including inherent limitations in the scientific method, our inadequate data base and research funding, the multivariate nature of the real world we are studying, and our present limited knowledge.

We know a good deal about how to handle the technical aspects of multivariate causation, including the possibility that sets of endogenous variables mutually influence one another through feedback mechanisms involving varying lag periods. By this I mean that the basic principles and theory construction and theory-testing strategies are reasonably well understood.

But the practical complications such complexities produce are considerable, particularly when combined with others, such as the costs and difficulties of obtaining measures of a high proporation of the variables we would like to include in our models. There are also some important side effects, including those that affect professional norms and the morale of individual researchers.

If there are many causal factors that affect any given variable, we may anticipate that when only two or three are introduced simultaneously, the result will be a small percentage of explained variance. Furthermore, as additional variables are included, the increments in explained variance will tend to become smaller and smaller, especially whenever the independent variables are moderately intercorrelated. Since this pattern is likely to hold for a variety of different combinations of independent variables, it becomes difficult to reach definitive conclusions regarding which ones to eliminate. The pitfalls of using stepwise regression programs in an atheoretical fashion are well known, but the problem is not so much one of technique as one of substance. Correlation matrices containing intercorrelations ranging in absolute value between 0 and .40 are very common in microlevel research in which persons are the units of analysis. Given the existence of unknown measurement errors and sampling fluctuations, such matrices usually result in inconclusive analyses.

Multiple causation is especially likely to result in disappointing results whenever we pin our hopes on a single explanatory variable, perhaps a so-called new one or one that has been neglected in the recent literature. The same is true in connection with the evaluation of social programs that have involved the manipulation of only one or two factors over a relatively brief period. The conclusion that a particular program does not work may nearly always be anticipated in advance given the complexity of the social causation. It will often be true that the proponents of the program in question have deliberately or subconsciously oversimplified the situation in order, as political advocates, to "sell" it to a legislative body. Regardless, the failure of the program to produce substantial results contributes to the more general conclusion that "nothing works." This is but a political manifestation of the more general point that single factors can hardly be expected to explain a high proportion of the total variance whenever the causal picture is highly complex. Nor is it realistic to pose two theories as simple alternatives to one another, presuming that the rejection of one suggests the acceptance of another.

Whenever one is dealing with a situation involving a relatively low percentage of explained variance, it is difficult to decide where the problem resides. Perhaps one has selected the wrong set of independent variables, and a more prudent search for others would have a significant payoff. The inadequacies may also stem from poor measurement and, in particular, from rather substantial random measurement errors that produce attenuations in parameter estimates. Or it may be the case that *no* small set of perfectly measured variables would yield a much higher explained variance since there may be upwards of 40 or 50 factors at work.

Given this ambiguity, what strategy can the investigator use? Following even one of the implied paths might entail a substantial effort over a long period, with very little hope for a major breakthrough. The temptation may be to give up prematurely. Such a tendency may then generalize to an entire field. For instance, the pattern of relatively weak correlations between single attitudes and behaviors that social psychologists typically found during the 1940s and 1950s seems to be one important reason why attitude research has declined rather dramatically in both sociology and psychology. In retrospect, one might ask why we should have expected a single attitude to predict well to any behavior, given the complexity of human motivation and the variety of constraining factors operating in diverse settings. Nevertheless, a persistent pattern of correlations in the .30 range seems to have had a dampening effect on this type of research in a variety of substantive areas.

This situation also suggests that there will be a large number of equally satisfactory alternative explanations for any given phenomenon. Each is plausible and accounts for just enough variance to resist its total elimination. The set of explanatory variables in one theoretical system is likely to be moderately correlated with those in another. In the case of any one or two explanatory variables, the associations with dependent variables will be small enough that it will be rather easy to suggest other variables that are possible sources of spuriousness.[1] If subsequent research tends to rule out this particular set of alternative explanatory factors, another can be conjured up rather easily. Disciplinary biases and division of labor will also enter the picture, with the result that there are likely to be several equally satisfactory clusters of explanatory variables that work to approximately the same degree. Social scientists are then free to choose among them according to their intellectual preferences.

As the number of intercorrelated independent variables increases, so does the number of reasonably plausible causal models that may be constructed to explain these relationships. There are six pairs of possible linkages among four intercorrelated variables, each possibly involving reciprocal causation. With 5 such variables there are 10 pairs, with 7 there are 21, and with 10 there are 45. Teasing out the reasons why such explanatory variables may be intercorrelated becomes a major theoretical task, although answers to important policy questions may depend heavily on the assumed causal connections among them. That is, the assessment of indirect effects depends on the correct specification of these interrelationships.[2] Yet if one simplifies by merely omitting those variables that are most difficult to measure or interpret, it becomes difficult to justify needed assumptions about covariances among disturbance terms.[3]

Another complication occurs whenever we allow for different models or temporal sequences for different actors. In the simplest of cases, X may precede Y for actor A, whereas the reverse may be true for actor B. For instance, one woman may elect to defer having children because of her position in the labor force, whereas a second may defer entering the labor force because of young children at home. If these temporal sequences are not recorded, then even though recursive models may be appropriate for the separate cases, one may need to allow for reciprocal causation for the pooled data. Homogeneity assumptions of some kind will always be necessary, and in regression analyses these usually take the form of assumed constant coefficients for each of the explanatory variables.[4] The more such coefficients there are, however, the greater the likelihood that at least some of these homogeneity assumptions will be invalid for a substantial portion of the population being studied. The more generally we wish our theories to hold, the more problematic such assumptions become. Yet to check on their adequacy will require additional data, as for example information about temporal sequences and lag periods for each actor.

Multiple causation also creates formidable problems whenever indirect measurement is necessary, a point that will be illustrated in more detail in the next section. When one measures a variable in terms of its assumed consequences—as is done in physics whenever one infers temperature by examining the height of a mercury column or infers a body's mass by means of pointer readings near the surface of the earth—accurate measurement depends on the constancy properties of certain physical substances (mercury, copper) and on relatively simple and nearly deterministic laws of nature.[5] It is obviously risky to infer discrimination on the basis of inequality levels or attitudes on the basis of behavioral responses. In effect, one must supplement whatever substantive theories one is testing with rather complex auxiliary measurement theories, and the two become confounded in many instances. The temptation becomes that of assuming the measurement problems away by simplifying these measurement theories to an unreasonable degree. This enables us to retain our sanity and some sense of progress but at the expense of realism. It also means that our measurement procedures will usually be very open to challenge and that problems of measurement comparability are sidestepped. One common temptation becomes that of relying on measurement by fiat—simply announcing that certain measures have been used because they were the only ones available.

One important indirect consequence of these and other complications is that we seldom, if ever, experience any dramatic breakthroughs in the social sciences. Many of us look for new explanatory variables, but these are likely to be very similar to those that have been used previously, though the labels we give them may have changed. We may search for "master" independent variables that are useful in explaining a variety of phenomena, but we almost always find either that their explanatory power is low or that it overlaps considerably with that of variables already in general use.

We do not discover many new things, though sometimes we may exaggerate our findings to make it appear as though we have done so. We may offer prizes and honors to our members, but these are usually awarded more on the basis of rather slowly accumulating products than on flashy new ones—although we are sometimes fooled into believing we have located the latter. It is much easier to exaggerate the importance of relationships when one is presenting a theory without data or when one is accounting ex post facto for historical patterns that have occurred in a small number of countries, a matter I shall explore in later chapters. Because explained variance is usually disappointingly small, so is the expectation for rapid success. How this affects the everyday behaviors of researchers is difficult to assess, but I presume it creates a kind of intellectual atmosphere different from those few scientific disciplines where breakthroughs are reasonably common.

One temptation in the face of this lack of dramatic results is to exaggerate the importance of small differences and to place the emphasis on statistical significance and probability levels rather than magnitudes of relationships and explained variance. This tendency seems especially pronounced in the reports of those psychologists and social psychologists who rely heavily on analysis of variance, F tests, and the reporting of group means but who often fail to report the numerical values of correlation ratios or intraclass correlations. In secondary reporting of the results of such experiments, the temptation is to mention that the investigators found "significant" differences between subjects exposed to different treatments, without at the same time indicating that within-treatment differences were large as compared to between-treatment differences.

Another indirect result of the fact that social scientists must allow for multivariate causation is that a reasonable degree of statistical sophistication is required, on the part of not only those doing the research but also those who read the reports. Obviously, one cannot take the time to explain these complexities in a brief, nontechnical empirical report. Therefore one is faced with the dilemma of either writing for a relatively sophisticated audience and losing one's other readers or of watering down the research report, if not the actual data analysis. In a very real sense a lack of methodological sophistication on the part of one's intended audience tends to reduce the report to the level of the least common denominator. It may also help to intensify any generational splits that happen to coincide with the sophistication of statistical and methodological training that was provided to different cohorts of social scientists. One may anticipate that if this level of sophistication has been changing at a rapid pace, those who have been left behind will adopt a defensive posture through which multivariate analyses are reflected altogether. In sociology and political science, for example, the "hard" versus "soft" cleavage has been with us for a long time but is only intensified by the rapid introduction of relatively sophisticated methodological tools of analysis.

The implications of multivariate causation are perhaps even more fundamental from the standpoint of data *collection*. If, for instance, 20 to 40 variables are required to provide an adequate explanation of some reasonably complex phenomenon, it is highly unlikely that nearly all of them can be collected by a single procedure—as, for example, a sample survey. Ordinarily data will have to be collected at several points in time and may involve two or more different levels of analysis. Community-level, neighborhood-level, and clique-level data may need to be combined with individual-level data in order to account for an individual's behavior. Observational data may need to be combined with responses to surveys. Multiple measures of the same variables may also require different data collection operations, as implied by the multitrait, multimethod approach originally suggested by Campbell and Fiske (1959).

Data are ordinarily available on a highly selective basis owing to practical constraints imposed at the data collection stage. Surveys simply cannot tap many kinds of variables, particularly contextual factors and network structures. Observational studies must necessarily be limited in time as well as settings. All of this has been recognized for a long time. The usual argument is then made that each data collection procedure needs to be complemented by others.

But how can one obtain all this information on each of the cases being studied? Certain compromises can be made provided that the design is well devised. By selecting subsamples of cases for which certain combinations of variables are measured, with the remaining combinations being covered by other subsamples, one can get around the necessity of obtaining all the pieces of information from a single respondent.

Clearly, this requires a high degree of coordination and advance planning. To the extent that there is a disciplinary division of labor with respect to the selection and measurement of independent variables, one can at least imagine such carefully coordinated data collection efforts being conducted by interdisciplinary teams using a diversity of research tools. Without such coordination and large-scale (and costly) data collection conducted over reasonably long periods of time, however, our existng data base will remain inadequate to test even reasonably complex causal models. Hangups of this nature have been especially important in blocking really serious efforts to conduct cross-level analyses in which micro- and macrolevel variables are combined into the same explanatory models.[6]

Obviously this problem is directly related to the costs of social research and the problem of allocating scarce resources. If we could agree on four or five fundamentally important areas of research promising important break-throughs that would shed light on a diversity of important theoretical and practical questions, we could coordinate our activities to encourage a small number of huge data-collection efforts, with the data being made available to individual social scientists for their own analyses.

A few models of such coordinated efforts do exist, two being the World Fertility Survey conducted in a number of developing countries and the National Election Studies surveys conducted by political scientists. These efforts have been facilitated by the fact that demographers are usually concerned with a relatively small number of fertility behaviors, as dependent variables, and political scientists are of course interested in voting behaviors and political attitudes during national elections, where the diversity of dependent variables is much greater, however—as in sociology—such highly coordinated efforts are few and far between. Needless to say, practical problems such as respondent fatigue, unwillingness to cooperate, residential mobility, inaccurate recall, and lack of question salience place additional restrictions on the data collection enterprise, regardless of the effectiveness of the coordination effort. Given all these problems, it is difficult even to figure out an efficient division of labor or a mechanism for getting such major research projects under way. So we do not make the needed effort.

INDIRECT MEASUREMENT, GENERALIZABILITY, AND MULTIPLE CAUSATION

Since the verification or testing of scientific theories relies on some kind of assessment of the goodness of fit between theoretical predictions and empirical evidence that must inevitably be filtered through our human sensory organs, all measurement must be indirect, though to varying degrees. This is not the place to enter into a lengthy discussion of the subtleties of measurement. Suffice to say that the pointer reading or other operational criteria that are used to provide cardinal, ordinal, or categorical information about specific events or properties must be linked in some fashion, either explicitly or implicitly, to the theoretical constructs or variables that enter our verbal discussions or formalistic theories.

Thus there must be some chain of theoretical reasoning involved in the measurement process. Sometimes this chain is relatively simple and the links are well understood, so much so that the measurement process is considered direct. Most of us, for example, would consider the measurement of the length of a table or the passage of time as sufficiently direct that we think very little about it. We would engage in serious debates, however, over the measurement of some human motivation such as "love," "altruism," or "competitiveness," or even about a more specific political attitude or form of behavior.

The highly precise indirect measurement obtainable in physics depends on two essential ingredients: some remarkable constancy properties of nature and a set of theoretical laws that are, for all practical purposes, both simple and virtually deterministic. Thus we infer and measure heat energy by examining the height of a column of mercury or the expansion rate of some other chemically pure substance with a known coefficient of expansion. Mass is inferred with the aid of physical measuring instruments that are assumed to retain constant properties during the period of their use, as well as gravitational force g that is assumed to be virtually constant near the earth's surface. Time is measured by taking advantage of periodicity properties of a pendulum or, say the earth's rotation or movement around the sun. A calibrated meter stick is assumed to maintain a constant length, or, if not, to be expanding or contracting in known ways that can be compensated for by applying deterministic correction factors. Very small unreliabilities are then assessed through repeated measurements, assuming that the property being measured is not undergoing indeterminate changes during this period of measurement. Whenever a given measuring device is incapable of producing a degree of reliability compatible with the precision needed, a substitute instrument is then developed and mathematical transformations made in order to assure comparability between the two instruments.

The causal laws connecting the physical phenomenon one is attempting to measure and the pointer readings that provide numerical values to be attached to that phenomenon are thus assumed to be simple, precise, and deterministic, even though at some micro level of analysis a certain degree of indeterminateness is presumed. Typically, one is using causal arguments that treat the phenomenon being measured as a cause of the indicator value, as for example the idea that an increase in heat energy *produces* an expansion of a metal or that the properties of a small particle produce certain traces in a cloud or bubble chamber. What if the expansion rate or bubble traces had multiple causes, only some of which could be controlled? Indirect measurement would become much less precise and the assumptions needed to justify it would also be subject to dispute.

This is precisely the problem we face in the social sciences, where our causal laws are both multivariate and indeterministic. Some simplifying assumptions will always be necessary, but we may not agree on which ones to use. Nor will all such assumptions be either obvious or explicitly brought to our attention. Measurement decisions then become much more problematic and seemingly arbitrary. It should come as no surprise, then, to find social scientists in sharp disagreements both about the measures of specific variables and also the more general problem of whether or not measurement issues should be seriously addressed at all.

An Example: The Measurement of Discrimination

Let us turn to an example that will illustrate a number of points about indirect measurement. Suppose we wish to measure the amount of employer discrimination in a number of different communities, perhaps with the goal of explaining differences in discrimination rates in terms of other community characteristics.[7] How is a behavior such as discrimination usually defined? Since discrimination, like many other behaviors, may take many different forms that are not readily identifiable in terms of their manifest similarities, the term is usually defined either by reference to some internal state, such as purpose or intent, or by reference to its supposed consequences.

A common ingredient in nearly all definitions of discrimination is the notion that there must be some *differential treatment* based on criteria that are deemed "irrelevant" (by someone) to the performance of some expected or actual behavior on the part of an actor who is being allocated some resource. For instance, suppose several persons are in competition for the position of quarterback on a football team. If some of these persons are rejected because they are over 50 years old, totally lack the coordination necessary for agile reactions and accurate passing, or are women or very young children, we do not consider this an act of discrimination. Why not? Because we recognize that certain characteristics improve the probability of satisfactory or superior performance and therefore may be used to evaluate the potential of the various candidates; they are "legitimate" criteria for differential treatment. We presume, however, that race is not a legitimate criterion in this instance. What about the selection of a salesperson or airline pilot, however? What if race or sex is correlated moderately with performance levels? Can it be used as a predictor and therefore a screening device?

Because of the fact that employers may rather easily use correlates of race or sex as selection criteria and then argue that these are legitimate in terms of differential treatment, we find it necessary to invoke some sort of causal argument that gets us into difficulty in the measurement process. Did the employer *intend* to discriminate by using, say, a battery of tests known to favor whites but also known to predict somewhat adequately to later performance on the job? What if women *do* have higher job turnover rates or slightly lower performance levels? How do we handle the legitimacy problem, given that nearly all predictors of performance are imperfect? Can we infer intent if our definition of discrimination calls for this? How would we obtain a measure of the *degree* of discrimination across a large number of employment decisions of this type? It should be noted, incidentally, that a similar problem of defining discrimination operationally has been faced by our legal system, where the issue arises as to whether the burden is to prove intent or whether the criterion should be defined in terms of results of discrimination.

Let us consider the option of defining discrimination in terms of its effects. If one assumes a single cause and a very simple world, all is well and good. The usual procedure is to infer discrimination on the basis of *inequalities* that may be more directly observed. If such inequalities were resultants of a single cause or if one could safely assume that all other causes of the inequality have been controlled, our problem would be solved. The measurement strategy here is one of "residualizing." One looks for inequalities, adjusting for all other *known* causes, and infers that these adjusted inequalities may then be equated with levels of discrimination. But who decides what these supposedly known causes are, and on what basis, and with what possible biases? Will the other causes, for which adjustments are to be made, be simply related to inequalities? If not, how should we handle them? This is a common kind of indirect measurement problem encountered in the social sciences.

Let us begin with the simplest possible situation depicted in Figure 1 in which inequality I is determined by only two factors, discrimination D and a residual or error term u, which we assume to be uncorrelated with D. If, in fact, D and u are uncorrelated—and we will never *know* this since D is unmeasured—then we may rather easily reduce the

Figure 1

variation in u by aggregating over a large number of acts and taking I as a measure of D that is subject to only relatively minor random disturbances. Unfortunately, this extremely simplistic model is often used to infer discrimination rates, which are in effect *equated* with inequality levels—say those between incomes of blacks and whites or men and women.

What if there are other causes of the inequalities, however? Let us pick just three: educational differentials E, inequalities in the job experience levels J of the applicants, and ability differences A. Suppose, furthermore, that the first two of these variables have been measured and that the ability factor is absorbed into the error term u. Our model might then look like that represented by Figure 2, in which we assume no particular causal connections between discrimination D and either educational differences E or job experience differences J. Now we would obviously want to control or adjust for both of our measured variables E and J, using the *adjusted* inequality levels as our revised measure of discrimination.[8]

This is, in fact, the most common type of measure of discrimination in current use. Notice that it requires the *simultaneous* controlling for *all* of the alternative causes of the dependent variable, except for those that we are willing

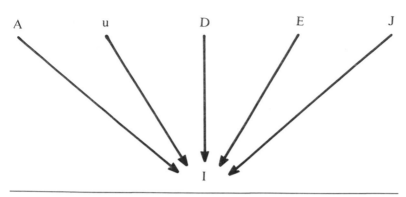

Figure 2

to place in the error term u. Having placed the unmeasured variable "ability" into this error term, what must we assume about it if we are to avoid a biased measure of discrimination? We must assume that it and all other neglected variables are uncorrelated with D, the unmeasured cause of inequality that we are attempting to measure by the residualizing operation. What if ability is correlated with the employer's behavior, however, perhaps because the employer has obtained some information unavailable to us? We must either assume the problem away by presuming that our two control variables, education and experience, have captured the ability differential, or we must obtain a (nearly perfect) measure of ability and bring it into the causal system as an additional control variable. Similarly, we assume that all other causes of the inequalities—including the decisions of applicants as to whether or not to accept the offers—have either effectively been held constant along with our explicit controls or are uncorrelated with the employer's behavior.

Now let us look at another sort of complication, represented by Figure 3, in which there is a presumed causal connection between the employer's discriminatory behavior

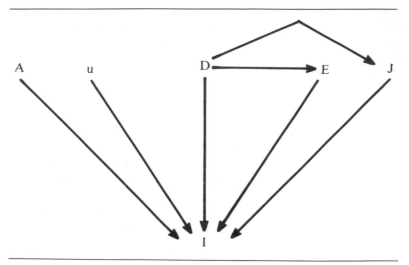

Figure 3

D and our two control variables, education and experience. Suppose, for example, that blacks have been prevented from getting the necessary experience as a result of previous acts of discrimination. Perhaps their educations were hampered or discouraged, partly as a result of the employer's actions, those of similar employers, or those of teachers who believed that black students should not be encouraged because they would later be doomed to disappointment. Should we now control for education and experience differentials? The picture is not so clear. We certainly would want to distinguish between present levels of discrimination—which presumably do not affect present educational or experience levels—and past levels of D, which may. This requires us to make some definite assumptions about the time lags involved. One possibility is diagrammed in Figure 4, in which discrimination at time t-1—a variable that also would have to be measured indirectly—influences discrimination at time t, as well as educational and experience differentials at the later time t.

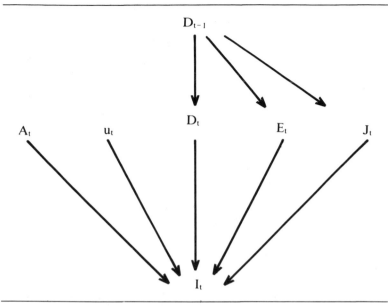

Figure 4

It is also likely that employer discrimination derives from attitudes that are based at least partly on certain aspects of the minority's past behavior, though those of us with liberal biases may be tempted to omit such a factor from our explanatory systems. Perhaps this could be represented by taking discrimination at time t as a function of educational differentials at time t-1, with these earlier differentials also affecting educational differentials at time t, perhaps through a generational effect. Such a model could be represented by Figure 5. Finally, it is also possible that discrimination at time t is affected by earlier levels of inequalities, say at time t-1. One mechanism would be through employer expectations, based on real experiences or secondary evidence, to the effect that minority performance levels will be inferior to those of majority members. Such a possibility is diagrammed in Figure 6, which is by now a far cry from the highly problematic models of Figures 1 and 2.

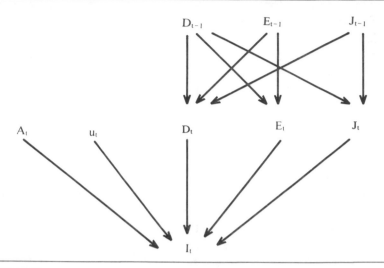

Figure 5

Far more complex and perhaps more realistic models could be constructed to allow for socialization differences, minority hesitation to apply for appropriate jobs in the first place, and measurement errors in any of the previously mentioned variables. For example, it is highly unlikely that a simple measure of years of formal schooling completed could allow for quality differences or differences in curricula or in ability to perform a complex job-related task. Measurement error models cannot become overly complex or the estimation process will become empirically hopeless. If there *is* a complex process at work that modifies a simple cause-and-effect relationship between employer discrimination and resulting inequalities, however, our auxiliary measurement theory must approximate this process if biases are to be avoided.

What are some of the more obvious implications of this exercise? First, we see that what may have started as a relatively straightforward problem of measuring a phenomenon in terms of its supposed effects is likely to be broadened into a theoretical problem of accounting for the complexly related causes of inequality. Thus theoretical and measurement issues are almost inextricably intertwined. One does not simply "go out and measure something" in order to do an empirical study.

Second, we see that indirect measurement requires a number of simultaneous assumptions, and shortcomings in any of them may produce serious flaws in the entire argument. The theory interrelating the several postulated causes (here of inequality) must be well specified, including the time lags, whenever we wish to allow for delayed reactions or feedbacks. There will always be assumptions required by this process, at least some of which will not be testable with the data at hand. For instance, we will have to assume something about omitted variables (such as ability differences) whose impacts are aggregated together into our error terms. We will also need to have high-quality data on many other variables in which we are not directly interested. If one's measures of "education" contain biases, if the appropriate data are missing for some of our cases, or if the available data apply to the wrong time periods, then there will be additional errors of unknown magnitudes.

Third, we must also admit the possibility that our measurement error theories—say that implied by our final model of Figure 6—may not be equally valid in all settings. Perhaps simpler models will not be misleading in some settings but would produce serious biases in others. As a general rule, the more diverse the settings and the more indirect the measurement, the more complex our measurement error theories must be. A procedure that works very well in a laboratory setting may not carry over easily into a more natural setting; nor can one count on observationally similar research operations being equally valid or reliable across such situations.[9] As a general rule, the more widely we wish our theories to apply, the more sensitized we must

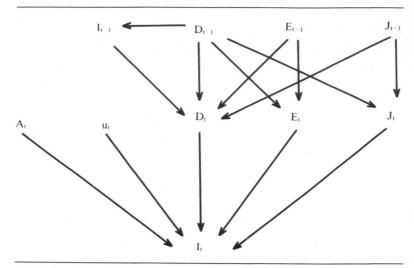

Figure 6

become to possible differences in measurement error models across these settings.

Finally, it is obvious that the measurement process may become highly technical. Simple rules of thumb, perhaps suggesting controls for one or two easily measured variables, must give way to much more complex statistical models and the use of computer programs that are capable of building in large numbers of complications simultaneously. The technical tools for handling such complications are now available, as are reasonably inexpensive computer programs for their use.[10] The mere existence of such technical tools is not sufficient, however. First, the assumptions underlying these tools must be well understood by practicing social scientists, and this requires a rather extensive period of training. Second, the necessary data must be available and at least some of the variables nearly perfectly measured. And, most difficult, our theories must be adequate to enable us to specify rather complex patterns of causation, including the reasons for the intercorrelations *among* the explanatory variables, both the ones we wish to measure and those that must be controlled. Without good measures in the first place, however, it is difficult to assess the adequacy of the measurement error models used to obtain these measures. In a very important sense we find ourselves going in circles.

We see in all of this an implied dilemma. Just how much does one simplify reality and proceed as though the answers provided are accurate ones? Here is where our intellectual and ideological biases are likely to enter in, in either a blatant or a more subtle way. One may, for instance, merely assert that inequalities are a measure of discrimination and ignore the complications. Or one can make a half-hearted effort, leaving out those complicating factors that might be expected to weaken one's argument and claiming that these variables could not be measured. If one equated blacks and whites on years of formal education, neglecting possible unmeasured differences in quality of education, several stances are still possible. One is to slur over the quality differences by simply stating that "education" has been controlled. Another is to admit that quality has been ignored but claim that there are no real differences or that quality differences will somehow cancel out. Another is to note the omission and suggest the direction of bias but then later ignore it in making policy recommendations. All of these options fall short of the scientific ideal but are real temptations, both to keep our intellectual sanity and to make policy recommendations that have any chance of being taken seriously.

NOTES

1. In a simple three-variable model in which Z is the source of a spurious correlation between X and Y, and which we may diagram as $X \leftarrow Z \rightarrow Y$, the correlation between X and Y will be less in magnitude than the associations between Z and both X and Y. Apart from sampling errors, r_{xy} will be approxi-

mately equal to $r_{xz}r_{yz}$. Thus if $r_{xy} = .30$, this may be explained by locating a variable Z such that $r_{xz} = .60$ and $r_{yz} = .50$. If $r_{xy} = .60$, however, it would be much more difficult to locate a *single* source of spuriousness.

2. For discussions of path analysis see Duncan (1966, 1975), Land (1969), and Namboodiri et al. (1975).

3. For a more extended discussion see Blalock (1982: chaps. 5 and 6).

4. In the simplest regression equation $Y = \alpha + \beta X + u$, for example, we assume that all cases are characterized by the same coefficients—that is, that α and β are really constants. Thus a change in X of one unit will produce an expected change of β units in Y for all individuals. More realistically, however, β (and α) will be a *variable* to be explained by other variables in the causal system. At some point, though, one must make *some* homogeneity assumptions to the effect that certain parameters are genuine constants. Such assumptions will be more or less realistic depending on the nature of the population and settings to which one is attempting to generalize.

5. See Krantz et al. (1971) for a definitive discussion of these constancy properties. See also Blalock (1982).

6. For reviews of the literature and issues relating to cross-level and contextual-effect analyses, see Boyd and Iversen (1979), Farkas (1974), and Hauser (1974).

7. This example of the measurement of discrimination is discussed in more detail in Blalock and Wilken (1979).

8. In adjusting, we also need to assume that E and J have been measured perfectly and that we know the forms of all equations. We usually assume the latter to be linear, sometimes without any supporting evidence.

9. This argument is developed more fully in Blalock (1982).

10. The most recent version of the LISREL program is provided in Jöreskog and Sörbom (1981). For an excellent expository discussion of the models and procedures, see Long (1976).

CHAPTER
3

Further Ambiguities
in the Real World

There are far too many kinds of complicating factors that confront the social sciences to warrant even nontechnical discussions of each within this single brief treatment. Therefore a sampling of issues becomes necessary. The following two chapters will be concerned with the problems one encounters in moving from a large number of facts to many fewer lawlike generalizations and to dilemmas encountered in the simplification process. In the present chapter, by contrast, I have selected two very different though pervasive kinds of problems encountered in each of the social sciences. First, I shall examine some of the implications of the fact that social reality is often fuzzy and apparently incapable of being described or measured in ways that are simultaneously precise and yet not misleading. In the second section I then turn to problems encountered in formulating dynamic theories that can be assessed empirically whenever real changes are both continuous and relatively small as compared with the magnitudes of our measurement errors.

LIMITATIONS IMPOSED BY
A FUZZY REALITY

We all know that ambiguities and unclear situations abound in the social world that surrounds us. Such ambiguities not only create problems for us as actors, but they also result in dilemmas for the social scientist trying to describe and explain social phenomena. In this section I focus primarily on a series of measurement-conceptualization problems related to such ambiguities, but the problem is far more general. In effect, very real limits are placed on measurement precision and therefore on our ability to verify or disconfirm any theories that are stated with a degree of precision that is unwarranted by virtue of this fuzziness that appears to be inherent in social reality. We appear to have an analogue to Heisenberg's Uncertainty Principle in physics, which places a limit on measurement precision owing to the nature of the physical reality being studied.

Most of our attention will focus on ambiguities in group boundaries, so it is perhaps advisable to discuss several other kinds of ambiguities that produce similar difficulties in order to suggest the generality of the problem. In sociology we are fond of discussing group norms, deviance from such norms, and the nature of the sanctions that are applied to deviant actors. Yet very few social norms are crystal clear. Sometimes we have idealized behavior proscriptions or prescriptions, as for example the Ten Commandments. We rarely expect anyone to live up to such norms completely, however, so that we typically find that the real or operative norms of a group are much more vague and slippery.

It therefore becomes almost impossible to give a precise meaning to the notion of "degree of deviance" from these elusive standards. In fact, a certain amount of tolerance of deviance itself becomes a norm, so that minor infractions of rules are hardly noticed. In effect, in most instances we are permitted to engage in a *range* of behaviors, being sanctioned only if we depart from this range by a considerable amount. Thus parents are expected to treat their children kindly, to protect them from injury, to help them mature, and so forth. Since all of these behaviors are subject to differing interpretations, however, and since expectations placed on parents are often somewhat incompatible, there is no precise set of expectations or behavioral acts on which we can all agree. Perhaps we can cite examples of some bad parents and some good ones, but there is a broad gray area within which parenting behavior is defined as "normal." The same obviously applies to teaching and to virtually all occupational performances, to what constitutes a good neighbor or church member, and—indeed—to almost all reasonably complex roles we perform.

Suppose we have formulated some propositions about deviance, the conditions under which it is more or less likely, and how it is sanctioned. How is deviance about imprecise norms to be measured? One possible line of attack is to concentrate almost entirely on norms that have been well defined—as, for example, those that have been embodied in our legal codes. In fact, much of the field of deviance in sociology is really a study of crime and delinquency rather than the myriad of other forms of deviance that are much harder to measure. When we adopt this tack, in effect we restrict ourselves to a small subset of deviant acts, merely hoping that findings that apply to this subset can be generalized to other forms.

Another approach is to select a plausible zero point and to measure degrees of deviance around this point, even though the true zero point is not a precise point at all, but a blurred region. Suppose that the social scientist considers suicides to be acts of deviance, defining the "deviance rate," in this instance, in terms of departures from zero suicides. It then becomes possible to compare deviance (of this particular form) across societies or communities in a precise way. But in almost all societies certain suicides will be defined as nondeviant acts. Someone who jumps into an icy river to save another may, in reality, be committing suicide, but he or she will be judged a hero. Elderly persons with fatal illnesses are rarely perceived as deviants if they commit suicides, but the extent that this is the case may vary from one setting to the next.

The point is that our measure of suicide rates can be treated as a valid measure of deviance only by making some rather strong assumptions about the precision with which norms have been defined and the similarity of these norms across settings. In effect, the apparent precision is specious. Put another way, we are measuring something (here suicide rates), but the something we are measuring is not what our theory is all about. Tests of the theory can therefore only be relatively crude. If we try to claim otherwise we are kidding ourselves.

Consider another kind of imprecise or slippery notion, that of a job candidate's "qualifications" or "potential." Suppose an employer claims to be evaluating job applicants in terms of either or both of these notions, with the decision of whether or not to hire being based on some imprecise subjective evaluation of each candidate. Suppose that we also wish to measure or assess the degree to which the employer is discriminating against, say, blacks or women. How do we proceed? First, we are likely to count the number of whites and blacks (or men and women) who have been hired, dividing these numbers by some other figures to standardize them. But what do we use for our denominators? The most common approach is to supply a precise criterion to replace the fuzzy ones actually used by the employer. For instance, we may take as our base figures the number of whites and blacks, within a given age range and living within a given geographic area, who have met certain minimal eligibility requirements—as for example, completing high school or graduating from college. We then consider all of these eligible candidates to have equal qualifications or potentials, or at least we assume identical frequency distributions for both groups. We may then obtain a proportion hired among blacks and compare it with the similar figure among whites, attributing inequalities to discrimination. Or we may standardize for an additional set of variables, using more sophisticated weighting schemes that we hope can serve as a standard by which to approximate the employer's much less precise method of assessment.

Perhaps our procedure is in fact superior, in the sense that it would provide a better method of predicting later performance. This is not the question, however, since the employer will undoubtedly invoke judgmental criteria that we will not be able to measure. The fact that our criteria are more precise than hers simply means that we have produced a different reality from the one that exists in the hiring situation. We are most certainly not reproducing it faithfully, and we will once more create a situation involving specious accuracy. The actual standards being used are far more imprecise than we have imagined, and our hypothetical employer will be quick to point this out.

Fuzzy Group Boundaries

Some groups such as business firms, universities, and nations have rather clear-cut boundaries such that one can readily decide who is and who is not a member. But many others do not, or at least the officially defined boundaries do not coincide with those that actors themselves define as real. Thus church membership lists are often padded with virtually inactive members. Most friendship groups have rather loose boundaries, with some persons easily classed as insiders but with others who are marginal and who will be classed as insiders by some and as outsiders by others. Although communities may have official political boundaries, we all recognize that these are arbitrarily defined. Furthermore, by historical accident some older cities, such as Boston, have much smaller political boundaries than many of our newer cities. For some purposes—such as voting and taxpaying—these arbitrarily drawn boundaries may be highly relevant, whereas for others—such as where we work, shop, and recreate—they are ignored. The boundaries of a school district are well defined but those of a child's peer group are not.

All of this is obvious, of course, but how do we deal with these situations when we need to obtain *measures* that we wish to attach to these fuzzily defined entities? How do we decide whether or not our measures are comparable from one group to the next when their own boundaries are unclear? Let us consider several examples. In sociology we like to stress the importance of "contextual effects" on individuals' behaviors. For instance, we believe that a child's behavior is influenced not only by his or her own characteristics but also by those of the family, peer group, and school setting. We would like to attach scores to these various contexts so as to assess the relative importance of individual- and group-level variables. Perhaps the scholarly orientation of a child's friends affects his or her learning. Maybe there is a school "climate" that can be estimated by taking school means on such things as parental income or classroom performance levels. We might even obtain attitudinal scores and average these across peers. But what unit do we use? The entire school? The student's classroom? The scores of students of the same sex? Scores of the student's five best friends? Most of us are influenced by those around us but to varying degrees and by different persons. For one child the relevant context may be a classroom, whereas to a second it may be a small group of close friends. There will be all sorts of overlapping groups. An older sibling may also be a close friend. Some adults may act and be perceived as though they were an additional set of peers, but this may vary across the children concerned. Some children have friends in other classes or as neighbors who attend other schools.

Unless we possessed a tremendous amount of information and almost limitless patience, as social scientists we obviously must impose certain arbitrary simplifications. But how? Often we act as though group boundaries are clear-cut, perhaps using a class average to measure the contextual variable appropriate to all members of that class. Or we may lack the information about each class and substitute an entire school mean on parental incomes. We then obtain a precise measure of something, but it is seldom the something we intend to measure. Furthermore, we must make additional assumptions to the effect that whatever errors we have introduced in one setting will be similar to those produced in another.

To take another example, suppose we wish to measure the degree of residential segregation in a number of different cities. Without going into details of the measurement process, suffice to say that all of the measures in common use require that each city be subdivided into small units such as census tracts or city blocks. The percentages of blacks and whites are then obtained for each of these subunits. There will be zero segregation when the minority percentage in every subunit exactly equals the minority percentage in the city itself. Complete segregation obtains when all of the blacks are in all-black subunits and all whites are in all-white subunits. The *degree* of segregation, however, depends on the distribution of nearly all-black, nearly equal, and nearly all-white subunits.[1]

When we recognize, first, that measures of segregation will depend on what subunit has been selected, whether these be of equal size or not, and on which of these subunits happen to be included within the arbitrary boundaries within which data are available, we realize that the choices of subunits and boundaries may be crucial to our results. For instance, in the case of black-white distributions, segregation scores tend to be much higher when city blocks are used as subunits, as contrasted with the larger census tracts.[2] Blocks, too, differ widely in terms of the number of persons who reside within them. In areas containing single-family units and duplexes, the block may constitute a realistic small-scale neighborhood, whereas a block containing a high-rise condominium or huge apartment complex may be far too large.

If a city's boundaries extend outward only as far as a mixed residential zone, whereas the suburbs are nearly entirely all-white, its segregation score may be deflated, as compared with another city that extends farther outward. To a limited extent the Census Bureau is able to reduce the distorting effects of such boundary problems by supplying some (but not all) data for what are called standard metropolitan statistical areas (SMSAs), but these areas, too, are defined in terms of county boundaries and arbitrary density criteria. In the case of data needed for the construction of segregation indices, block data are generally available only within central cities defined by political boundaries

As a final example of problems arising when boundaries are fuzzily or arbitrarily defined, consider the notion of "society," as defined in the anthropological literature. Modern industrialized societies are typically defined as nation-states, with well-defined (though often disputed) political boundaries. But how about Northwest Coast American Indians or Bantu "societies" prior to their domination by white colonial powers? How many were there and how would they be defined? One may use linguistic criteria and obtain one approximate answer. But if one studies the political organizations or trade networks among such groups, it is not clear where one draws the line. If each society is taken as a separate case for statistical analyses, what is our sample or population size? The answer is likely to depend on the criteria used, and the boundaries defined in terms of different criteria will inevitably overlap.

Not only do we again encounter measurement problems, say in designating the size of each such "society," but there will be important inference problems as well. For instance, one is often interested in treating each case as an independent replication, so as to assess whether an association between two variables or characteristics is likely to represent some lawlike connection between the two variables or whether it might have occurred simply by chance.

Suppose we find that societies at the hunting and gathering level of technology are characterized by an absence of slavery or by very low levels of economic inequality, whereas those at an advanced horticultural level are more likely to utilize slaves and display considerable inequalities in wealth. If this occurs regularly in 50 societies, the association is probably not coincidental. Or is it? Suppose nearly all of the hunting and gathering societies are spaced close together and share similar linguistic patterns. We would presume that many cultural patterns would then spread by diffusion, so that in effect there is only a single "replication." If the advanced horticultural societies are also close together, perhaps slavery was introduced into one of them and then spread to the others. Then we would not have such a large number of mutually independent replications. If we know that one society has slaves, this will enable us to predict, better than chance, that an adjacent society will also have them. In fact, someone could easily say that we are really not dealing with distinct societies at all, but rather a number of groupings within a single society or perhaps four or five societies.

The same kind of argument can be applied to spatially adjacent units of any type, say census tracts within a city or counties within a state. Two adjacent counties will look very much alike, often sharing common industries, labor forces, newspapers and TV stations, and persons who commute back and forth between them. How many distinct "regions" do we have within the United States? How many "subcultures" within a city? Or "neighborhoods"?

I do not wish to suggest that boundary problems such as these create insurmountable problems, since there are technical devices that may be used to assess the degrees of seriousness of distortions they may produce. For instance, diffusion across adjacent boundaries may be assessed by looking at the patterns of residuals or errors to see whether these are related to spatial locations.[3] A more general approach is to slice the cake in a number of different ways to see whether one's choice of boundaries really makes a difference in the empirical results. In our segregation example one might delete a number of blocks near a city's boundaries and recompute the measure of segregation, or one might combine blocks into pairs and triads, obtaining somewhat larger subunits.

Sensitivity analyses of this nature are in effect empirical reliability checks that address this question: Suppose we do not know the appropriate boundary and must make an arbitrary decision; how sensitive is our measure to the particular decision we make? If the results differ very little, regardless of our choice, then there is at least some justification for going ahead. But if they do differ considerably, a note of caution has been introduced. Either the boundary must be defined to coincide with the real one—if this is indeed a meaningful statement—or we must admit to an indeterminacy. We hope, of course, that our sensitivity analysis will give us a realistic estimate of how substantial this indeterminacy is, as well as the implications it has for linking whatever variables we are trying to measure to those that are not so dependent on the boundaries that have been selected.

COMPLICATIONS IN DYNAMIC ANALYSES

Nearly all social scientists agree on the desirability of conducting longitudinal as well as comparative studies and of being able to construct dynamic theories capable of explaining and predicting change. Having the objective is not enough if the many complications in dynamic analyses are not faced squarely, however. In particular, the effects of substantial measurement errors and data collection constraints place truly major restrictions on the study of change processes and may, indeed, imply that greater weight in many instances should be given to static, comparative analyses. For both comparative and longitudinal analyses, of course, one must be aware of the major sources of erroneous conclusions so that potential biases can be estimated and the implications of hidden assumptions brought into the open. In this section I shall focus nontechnically on some of the major complications that commonly arise in handling change data and in making inferences concerning the dynamic processes that underlie them.

One major source of difficulty arises because of the combination of measurement errors and the typically small magnitudes of real changes that can be studied during a brief time span. For instance, strictly random measurement errors attenuate relationships in proportion to the ratio of the measurement error variance to the variance in true scores.[4] This means that unless the changes in true scores are substantial, it becomes difficult to distinguish true change from random measurement error. Since in the multivariate case there will also be differential measurement errors in the several independent variables, which will also be intercorrelated, efforts to estimate coefficients will be especially handicapped by even relatively small measurement errors. Given that the differences that often exist *among* distinct individuals are likely to be much greater than the changes *within* these individuals, change studies are often much more sensitive to measurement error distortions than are comparative studies. This need not be the case, of course, especially when long-term changes are being examined. But for such long-term changes, one then encounters many of the same problems that characterize comparative studies: lack of comparability of measures and lack of control over extraneous factors that one is willing to assume are virtually constant during a briefer interval.

The problem of autocorrelated disturbances is a familiar one to students of time-series data, and a number of procedures have been discussed in that body of literature for postulating the mechanisms producing such autocorrelation, testing for its presence and removing its effects by working with difference functions.[5] What I have not seen well handled, however, is the *simultaneous* treatment of autocorrelation and measurement errors. Indeed, the procedures commonly recommended—namely, working with difference functions—seem hypersensitive to measurement errors, implying a tradeoff through which autocorrelation problems are exchanged for measurement error sensitivities.

For instance, suppose the disturbance at time t is taken as a linear function of the disturbance at t-1 (only) and a nonautocorrelated term, as in the equation

$$u_t = \rho u_{t-1} + \eta_t$$

where η_t represents the nonautocorrelated component and is a positive coefficient less than or equal to unity. One may eliminate the autocorrelation by taking difference functions for each of the variables—for example, $X_t - \rho X_{t-1}$. But unless measurement errors in X are negligible, they are likely to dominate the difference term $X_t - \rho X_{t-1}$, especially when ρ is close to unity. This is not a necessary result of the differencing operation, of course, but it is likely to be empirically the case over a brief time period. This is completely aside from the question of whether the actual changes are lasting ones or important in a theoretical sense.

Another very important practical concern is that of data availability, especially as one moves backward in time or must rely on retrospective reports of respondents. Given the greater data gaps that often exist for earlier time periods, as well as the differential adequacy of measures across periods, what are the temptations for the analyst? The first is to pretend that measurement errors do not exist or that they are constant across the period under study. In the case of strictly random measurement errors, this latter assumption would amount to assuming that the ratios of measurement error variances to variances in true values at time t are exactly the same as at t-1, and so on. Whether such simplifying assumptions are any more realistic in change studies than comparative ones will, of course, depend on the circumstances. But if true change scores are smaller in magnitude than are differences among true scores in a comparative study, then the former type of design is more sensitive to invalid measurement error assumptions of this type.

A second temptation, also a common one, is to assume that variables that have not been measured are actually unimportant and can safely be ignored. Any specification errors produced by such simplifying assumptions are thereby neglected. Event-history[6] techniques that are currently making headway in sociology, for example, rely heavily on asking respondents to recall the dates of significant events such as job changes, births and deaths of family members, spatial movements, graduation dates, and so forth. This permits the tracking of sequencing patterns that may be unique to each individual and that may be important determinants of subsequent behaviors. Many other changes occur gradually, however, without any conscious awareness of their taking place or at least without being able to pinpoint when they have occurred.

If such changes are simply left out of the analysis because the analyst rightly suspects that accurate recall would be exceedingly difficult, how will this affect the analysis? There is no way to tell. The temptation may then be to assume that dramatic events, rather than gradual shifts, are the really important causal factors accounting for one's dependent variable. This may or may not be the case and may depend on the phenomenon one is attempting to explain as well as the particular respondents being studied. Where the time period is substantial and the data gaps obvious—a problem familiar to archeologists and historians—one may be far more cautious in inferring causes than in instances where one is provided with supposedly "rich" event-history data.

Distributed-lag models used by econometricians are often realistic tools for handling continuous changes in instances where one's observations necessarily take place at discrete intervals. For example, if one believes that an actor's behavior Y_t at time t depends not only on X at time t but on earlier periods as well, one may write an equation of the form

$$Y_t = \alpha + \beta_0 X_t + \beta_1 X_{t-1} + \beta_2 X_{t-2} + \ldots + \beta_k X_{t-k} + u_t$$

where the time periods t, t−1, t−2, and so forth may be arbitrarily small. Even if one's data points involve much longer time intervals—say every kth smaller interval—one may still estimate the coefficients provided certain lawlike relationships are assumed to hold among the betas. For instance, if one assumes that the magnitudes of the betas decay exponentially over time, a smaller number of coefficients may be used to represent the model and estimates of these parameters may then be obtained with data collected at only a few points in time.[7]

The practicality of models of this type, however, depends once more on one's willingness to assume that measurement errors are either negligible or of certain restricted types. If the quality of one's data deteriorates as one goes back in time, as will almost inevitably be the case for retrospective data, then there will be unknown differential biases in one's estimates. If the relative variances in random measurement errors increase as one goes back in time, then the effects of the more recent values will be overestimated. Multicollinearity in the values of independent variables across time will aggravate the problem. The shorter the time intervals—say between t and t−1—the more likely that X_t and X_{t-1} will be highly correlated and the greater the sensitivity of our estimates will be to very slight measurement errors in X. If one's data have been collected by others, if the procedures have been inadequately documented, or if measurement is highly indirect and auxiliary measurement theories complex, the temptation may be to close one's eyes to such problems. In particular, the absence of a high correlation between the *measured* values of X at times t and t−1 may be taken as evidence that the true values are also not highly correlated. In effect there may be no allowance for unreliability in assessing true multicollinearity.

Another temptation may be to impose homogeneity assumptions with respect to lag periods. Rates of change and the relative magnitudes of the beta coefficients may not be the same for all actors. Some such homogeneity assumptions must always be imposed if one is to avoid an empirically hopeless situation. Presumably, however, one would want to have evidence based on prior studies as to the degree to which individuals vary with respect to appropriate time periods and relative magnitudes of coefficients. Armed with such preliminary data, one could then conduct sensitivity analyses using Monte Carlo simulations to assess the seriousness of distortions produced by simplifying assumptions of a given type. The essential point is that time-series models require a good deal of evidence and assumptions, just as do their comparative counterparts. The growing body of literature on causal inferences from panel data has helped to pinpoint the difficulties encountered whenever one allows, simultaneously, for indeterminate lag periods and measurement errors.[8]

Careful specification of conceptual variables is also often necessary to avoid misleading simplifying assumptions about temporal sequences, even where cross-sectional data are being used. It is often presumed that since sex and race are determined at birth, and since formal education can be dated, that the time-order of such background variables can be determined rather easily. This is indeed the case, but if these are not the variables in which one is really interested but are merely serving as indicators or proxies for experience variables, the dating problem is no longer straightforward. If females or blacks are continually exposed to situations that males or whites do not experience, or experience to lesser degrees, how can this be handled in a dynamic model? Some experiences may occur very early in life, others at rather specific times, and still others continuously or perhaps even cyclically. Certainly there are many different meanings and consequences of "education," not all of which can easily be dated. Variability across actors is again highly likely, in which case homogeneity assumptions about the coefficients need to be made with caution.

This brief listing of complications in change studies is not intended to suggest that difficulties encountered in such studies are any more severe than those often cited in connection with comparative, static research. They are sobering, however, and imply the need not only for considerably more methodological studies but for a concerted effort to collect better and more standardized data. If, somehow, social scientists can manage to gain greater control over the data collection process, we stand a reasonable chance of making steady progress. Economists have been fortunate to have much of their time-series data collected by others who have highly pragmatic interests in measures of specific types, especially outcome measures. A relatively selective set of variables are also measured by governmental census-taking operations, and some of these have been standardized across national boundaries. The social indicator movement has advocated the systematic collection of certain socially important measures that will enable us to evaluate progress, or the lack of it, with respect to certain noneconomic variables such as levels of satisfaction, discrimination and racial segregation, and physical and mental health.

As long as data are collected by others, however, these others will also be the ones who decide which variables to include and which to exclude. As already implied, one major tendency has been that of oversampling *dependent* variables—especially those of a rather dramatic interest, as for example crime and health statistics. This enables officials to keep track of trends, but it may be virtually useless in *explaining* such trends unless independent variables are also measured. The fact that this is an obvious point does not mean that it will be equally obvious to those who control the data collection process. The simple fact may be that those who collect the data may not *want* good explanatory models, either because they define their tasks simply or because the policymakers who pay their salaries may not wish to consider certain kinds of explanations or provide the necessary data to those who would like to explore them.

We also recognize a number of additional practical difficulties, not the least of which is the lack of rewards for long-term data collection efforts. Individual rewards must occur within a few years of data collection, especially where promotions and job tenure are at stake. Therefore lengthy time-series data must be collected by research organizations that need to find ways to reward their employees, other than by holding out possibilities of quick publications. Those who come along at a later time—say the fourth or fifth time period—will be in a far better position to extract meaningful results than those who made the initial effort, a fact that is obviously unfair from the standpoint of the individual researcher. This is completely aside from problems involved in keeping track of respondents and enticing them to continue cooperating in, say, a panel design. It also presupposes the resolution of difficult problems of lack of comparability of measurement over time, artifacts produced by repeated measurements, and the impacts of uncontrolled extraneous factors.

The basic difficulty is that the problems of collecting adequate time-series data seem so formidable that, collectively, we delay the moment for seriously considering how to organize to undertake the task. We must obviously be selective given the costs involved, and someone must be committed to continued financing. Few of us are so daring as to begin laying the groundwork and so the enterprise never gets started. We have seen several excellent beginnings, especially in well-financed areas such as consumer behavior research, election studies, follow-up studies on students, and in connection with selected types of crime. These popular areas may not be the most crucial in supplying the kinds of information that are most useful from the standpoint of developing more general theories, however.

If we make the optimistic assumption that the more readily funded projects are also automatically the most valuable in providing insights into more general social processes, we may only delay the major effort that seems needed if we are to gain greater control over the kinds of data we, ourselves, wish to collect. In short, the less organized and attentive we are to potential selectivity problems in this connection, the more our choices will be determined by factors that are extraneous to the goal of developing a cumulative knowledge base.

NOTES

1. For a detailed discussion of the measurement of residential segregation, see Taeuber and Taeuber (1965: Appendix A). See also Duncan and Duncan (1955).

2. Taeuber and Taeuber (1965: 220-231).

3. In the anthropological literature this problem of diffusion versus independent invention is referred to as "Galton's problem." See Naroll (1970) and Loftin (1972). This issue of spatial autocorrelation is analogous to but more complex than that of temporal autocorrelation, in that the former is two-dimensional and the latter unidimensional.

4. If the measured value $X' = X + e$, where X is the true value and e is a random measurement error component uncorrelated with X, then

$$\sigma_{X'}^2 = \sigma_X^2 + \sigma_e^2.$$

This increased variance then attenuates the estimate of β in the equation $Y = \alpha + \beta X + u$ according to the approximate asymptotic (large sample) formula

$$E(b_{YX'}) = \frac{\beta}{1 + \sigma_e^2 / \sigma_X^2}$$

Thus it is the variance in the measurement error component e *relative* to that in the true X that determines the attenuation in the slope estimate (or in a correlation coefficient).

5. For discussions of time-series procedures, see Box and Jenkins (1976) and Nelson (1973).

6. See Tuma et al. (1979).

7. See Christ (1966: 204-208).

8. There is now an extensive literature on panel data and measurement errors. See, for example, Duncan (1969), Heise (1970), Shingles (1976), and Greenberg and Kessler (1982).

Can We Move from Many Facts to Fewer Lawlike Propositions?

Social facts can be interesting and may stir us to action. But when they are too numerous or concern matters we know and care little about, it is often difficult to decide what to do about them. Indeed, they are likely to appear useless and hardly worth detailing. Anyone attempting to memorize such facts would undoubtedly find the experience mind cluttering, so much so that intelligent action would become almost impossible. How can such facts be processed and used to provide reality checks on our thought processes? Simplifications must be introduced. But which ones, how, and with what rationale? Given a high degree of initial complexity and detail, we would hardly expect most such simplifying schemes to lead in the same directions. But if they do not, how would one choose among them, and with what consequences?

Let us begin with some hypothetical "facts" that are strictly fictitious. Suppose we are given the following series of increasingly complex factual statements:

(1) In Seattle, in 1980, there were 600 divorces for every 1,000 marriages.

(2) In Seattle, in 1980, there were 700 divorces for every 1,000 marriages among whites, but only 400 divorces for every 1,000 marriages among blacks.

(3) For the same period in Tacoma, the white and black divorce figures were 650 and 450, respectively, indicating that the differential between white and black rates was smaller in Tacoma than in Seattle.

(4) In 1970, however, the Seattle figures were 600 and 500, respectively, whereas in Tacoma they were 500 and 450, respectively, indicating that the racial differential between Seattle and Tacoma was smaller in 1970 than in 1980.

What kinds of responses are likely in connection with the first statement? If one lives in Seattle, the reaction might be one of concern, depending on an implicit standard of some kind. These divorce rates are too high; we must do something about them! A resident of Boston might react with some surprise but pass the results off as being of little interest unless they were thought approximately applicable to that city. A resident of Australia, Japan, or South Africa might say, "I told you so; those Americans are in real trouble." This assumes, of course, that the persons concerned knew where Seattle is.

This simple example illustrates an important point. Facts about large and important places, or those close at hand, are more likely to attract our attention than those about lesser ones. We may have considerable interest in the facts regarding the rise of Adolph Hitler, but far less so if they pertain to similar processes in a remote African or Central American country. In the latter instances such facts would be of much greater interest if they could be used to confirm or refute a theory about totalitarian governments, whereas in the former instance we are far more likely to become fixated on the facts themselves. Facts about Kathy Cook or Bill Baker in East Cupcake, Kansas will be of little interest to anyone except to their neighbors. In general, we are seldom interested in facts about individual respondents in social surveys, not only because these persons are not known to us but because we do not want to become too overloaded with useless information.

What about the second of these factual statements? Here one would naturally be inclined to ask why there should be a difference between blacks and whites. Statement 3 poses a still more complex problem: Why should the black-white differential in divorce rates not be the same in Seattle and Tacoma? And in the case of the fourth set of facts, we are left wondering why the city-by-race differential is itself changing over time. We are almost automatically steered to a series of somewhat more abstract questions. Furthermore, if we were given comparable figures for each of 100 different cities, both Seattle and Tacoma would pass from view unless we were specifically interested in comparing them, blotting out all other information. We would certainly not want to go about memorizing all these facts.

Returning for the moment to statement 2, we would undoubtedly ask: What is it about the experiences or environments of black and white couples that produces a different divorce rate, at least in Seattle in 1980? We would now be in the process of attempting a historical explanation for a particular set of facts and would invariably invoke either an implicit or an explicit theory of some sort. This theory would contain a number of assumptions, many of which would be untestable with the data at hand. Perhaps divorce rates have something to do with average age at marriage, income levels, the educational differential between the spouses, the presence or absence of children of various ages, unemployment levels, or even the quality of the drinking water. Blacks and whites may have different average levels on one or more of these variables. Thus it may not be the biological factor of race but another factor that causes the difference.

Furthermore, Seattle and Tacoma blacks and whites may also have different levels on one or more of these variables, thus accounting for the city-by-race differential and, possibly, the temporal differences as well. Fine. There would appear to be many such causal factors at work, however, and with only two cities and two time points, how could we answer our questions? We could not. Thus we would be in the unfortunate position of not being able to choose among a wide variety of rival explanations for the presumed facts. We might prefer an argument that stressed the unemployment factor, particularly if we do not like high unemployment levels for other reasons. Or if we do not like whites, or if we prefer whites to blacks and also like high divorce rates, we could invoke a racial argument. Or we might blame the mayors of the two cities or the drinking water. Why not? Disputes over social facts are rarely this absurd, but the points readily generalize. First, we often lack the empirical data to resolve what would otherwise be reasonable alternative explanations. In this instance, for example, additional pieces of information collected on 100-200 cities would go a long way toward eliminating or rendering highly implausible at least some of the possible explanations.

Second, we readily see that as soon as one admits the possibility of multiple causation—perhaps as many as 50 factors that influence divorce rates—it becomes totally unrealistic to invoke the phrases "controlling for all relevant variables" or "other things being equal." They will not be equal and cannot all be controlled or even imperfectly measured.

Third, we recognize that in some instances there may not be enough cases—here, cities—to resolve the problem, even where all of the relevant variables are known and measured. Seattle and Tacoma, in our two-case example, differ in many ways. I can claim that the difference is due to the fact that Seattle is north of Tacoma. You could argue that it is because Seattle is larger or has a different unemployment rate. Since Seattle is both north of Tacoma and larger, how would we decide on the preferred explanation? How could we explain differences among England, France, Germany, the USSR, and the United States? There are far too many variables or dimensions along which they differ relative to the number of cases under observation. I shall return to this point after trying to state the issues involved in a more general fashion.

HISTORICAL FACTS AND SYSTEMATIC THEORY

In view of the tremendous complexities in the social realm, the impossibility of studying "everything," and the real practical limitations placed on data collection and thus the huge gaps in our factual knowledge, it is no wonder that social scientists are in sharp disagreement as to how best to proceed. In this section we shall examine one important kind of dispute that has existed in many specific forms and that relates closely to the question of how we allocate our scarce intellectual resources. Is it wisest to try to obtain detailed factual information about a small number of important cases, or should we go after fewer details but place a greater effort in the direction of raising the level of abstraction so as to construct systematic theories capable of predicting to selected events?

The issue can be seen in its purest form in terms of debates between historians or historically inclined social scientists on the one hand and science-oriented ones on the other. A historian will generally place a premium on gaining detailed factual information about a limited phenomenon, defined in terms of both space and time. Such facts will be meticulously checked for consistency with other pieces of information, at least to the extent of their availability. Facts that are "lost in history" may be inferred, but—at least in the ideal—speculations as to what such information might have been will be clearly distinguished from "reliable" sources. Much of the same kind of orientation exists among ethnographers, except that anthropologists can be firsthand witnesses of events, at least during a brief interval, and may also cross-check their information with live informants.

Why do we want such detailed information? Presumably it is because of the complexities involved and the desirability of gaining a complete explanation of the phenomena in question. Someone who lacks such details cannot hope to gain such a degree of understanding. But what do we mean by "understanding," and what is it the expert knows that the rest of us do not? We perhaps obtain a clue to such questions when area specialists are called on to explain, for instance, what is going on in Iran, why the Argentinian government decided to invade the Falklands, or why Israel invaded Lebanon at a particular point in time. They are being asked, ex post facto, to account for the behaviors of specific actors on the basis of factual information about these actors and the settings in which they are embedded.

Historians provide us with interpretations for wars, peasant uprisings, the collapse of the Roman Empire, and so forth. How can they do so without invoking a series of assumptions about the motivations of the actors concerned and the factors that impinged on them? In what sense does their understanding differ from that of, say, the family members of a man who has just committed suicide or the coworkers of a foreman who cannot get along with them? I can see no basic differences here, other than the obvious fact that only a few of us care about this particular foreman or the suicide victim, whereas it may be far more important for us to understand the Arab-Israeli conflict. We recognize that in all these instances it is far easier to give a *plausible* ex post facto explanation for the events in question than it is to predict them accurately in advance. This is because of the large number of variables that are potentially operative in each specific situation, making it easy to select plausible candidates from among them.

Suppose that the expert is also acquainted in less detail with four or five other cases—for example, several other countries, suicide victims or foremen. It then becomes possible to make a series of implicit comparisons and to develop working theories or crude causal models as explanatory mechanisms. I argue that this is precisely what must go on whenever an explanation for a series of facts is being attempted. Sometimes the comparisons are made explicit—for example, when one recites hostile acts among other Middle Eastern nations. Certain assumptions about human motivations must inevitably be invoked. Peasants behaved as they did because they wanted to be assured that they would not starve, resented being pressed into military service, or needed to find scapegoats for failing crops or an inexplicable epidemic. We often infer these human motives by appealing to our own experiences; or if we are aware of other sets of persons placed in similar environments, we presume they were similarly motivated. We may be willing to believe the assumptions made by an earlier chronicler of these events, presumably because of his (rarely her) greater familiarity with these events.

What I am asserting is the perhaps obvious point that explanations of *particular* events require appeals, no matter how implicit, to other events deemed in some way similar or comparable to those being explained. Thus one finds it necessary to get beyond the single case, and this can open the floodgates to a wave of difficult questions. Just what assumptions about the actors' motivations are being made, and how can they be justified? How does one decide whether several situations are sufficiently similar or comparable? Why should one believe a particular set of informants or authors, especially when one recognizes biases among one's own contemporaries? What evidence is there to support the implicit or explicit generalizations that often accompany explanations, say, for the invasion of one nation by another?

One of the things that statistical studies of contemporary events show us clearly is that explanatory variables suggested by common sense do not have anywhere near the predictive power we initially expect. Negative findings of essentially no relationship are common, as are those that yield correlations of the order of .20 or .30 and explained variances of 5%-10%. If this is true when we are looking at several hundred cases or more, why should we have much faith in relatively simple explanations for a single event? The reason that such explanations are appealing, I suspect, is that the historian is able to *select,* ex post facto, from among a large number of possible causes a relatively small number that have intuitive appeal for that particular event. If, for instance, two or three changes occurred prior to some major event—say the onset of war or an internal rebellion—these can be singled out easily. To account for a similar event in a second setting, one may also select out a few plausible factors, including some of the same ones used to account for the first one. Prediction of these events *in advance* would have been far more difficult, however.

We know from elementary algebra and geometry that if we have only two points we may fit a straight line through these points without error. In three-dimensional space we may fit a plane perfectly through any three points. More generally, if we have N cases and a total of N variables (N − 1 being independent and 1 dependent) we may explain our dependent variable perfectly by using a linear function of the N − 1 independent variables. Therefore, if we are examining even as many as 20 nations but have 19 independent variables, we know in advance that we will be able to explain 100% of the variance *regardless* of the data or facts.

What this says is that when a large number of factors are at work, we need to employ multivariate techniques to infer what is going on, and we must also have a sufficient number of cases to provide a fair test of our theories. If not, we can always select out the factors that work in the right combinations to account completely for the facts. Put

another way, if we have only a small number of observations, we cannot hope to have sufficient information to
falsify a reasonably complex theory. If we hold to one-
factor causation, our problem is simple enough, but when
we admit to multiple causation we ust have more evidence.
This simple point implies a great deal, however, in terms of
design strategies and the ways in which our data must be
collected.

The difficulty is not an intuitively obvious one, since we
often believe that our understanding has been validated
whenever it is satisfying to a substantial number of others,
who may share our intellectual biases, or whenever a
correct prediction based on this understanding has taken
place. There are several slippery problems here, one of
which is that often such predictions are not very specific.
"There will be a bloodbath in South Africa" is a prediction
many of us have made for a long time, and yet it has not
occurred. If someday it does, the "prediction" will have
been verified. If it does not, we can always wait a few more
years. We may also make predictions—without a theory or
understanding of the causal processes at work—simply by
extrapolating past trends or predicting that past behaviors
will be repeated.

If such forecasts are accompanied by some explanation
that seems plausible to one's readers, the whole intellectual
package may be accepted as constituting a prediction based
on theory. One suspects that this is precisely the kind of
knowledge possessed by experts on some country that is
unfamiliar to most of us. Such experts know the country's
past history and many current facts that may continue to be
relevant in the near future. They are also armed with a
selected number of causal explanations that may be
advanced, ex post facto, to account for some recent
dramatic incident. In addition, they may be able to project
recent trends into the near future, something those unaware

of such trends are unable to do. If they make errors, the chances are that we will not remember them because our attention will have been turned to something else, perhaps another dramatic event elsewhere on the globe.

This question of prediction is crucial to understanding the dilemmas involved; therefore let us examine it more closely. Suppose that an all-seeing mind is able to assure us that a dependent variable Y will vary in accord with a simple causal law of the following linear form:

$$Y = \alpha + \beta_1 X_1 + \beta_2 X_2 + \beta_3 X_3 + u$$

where α and the β_i are (for the time being) constants, and where the disturbance term u is responsible for 20% of the variance in Y but is totally uncorrelated with the independent variables X_i.

Interpreted causally, such an equation tells us that if X_1 were to change by one unit, and if X_2 and X_3 were held constant, then the *expected* change in Y would be β_1 units in the Y variable. This means that over the long run, changes in Y would average out to this amount. This is the kind of multivariate equation we usually attempt to estimate in the social sciences, though with fallible data and relatively little knowledge as to which variables to insert in the roles of the Xs. Setting aside any complications arising from this lack of knowledge, as well as our ignorance as to omitted variables that are implicit in the disturbance term, what can we say about how such an equation could be used to help make predictions about future values of Y?

Obviously, we would need to know the correct values of the β coefficients and to make the strong assumption that these βs are really constants that do not depend on the peculiarities of the cases we have selected. Not only must we know the present X values for our cases, but we must also be able to predict how *they* will change if we are to anticipate changes in Y. But perhaps these supposedly independent variables are not just correlated with one another, as is usual in nonexperimental settings, but perhaps some of them would change as a *result* of changes in the others. Maybe X_3 depends on X_1 and X_2, so that it is unrealistic to think in terms of X_1 changing while X_3 remains constant. Finally, there will always be sources of error produced by factors whose impact is summed up in the disturbance term u. Suppose some of these, too, depend on one or more of the Xs in the equation. Or perhaps some of them are causes of these Xs.

In the real world, of course, our equations would have to consist of many more than three independent variables, and our estimation problems would be compounded by measurement errors, high intercorrelations among the Xs, and many other complications. Thus it requires a good theory to make accurate predictions *about individual cases*. Furthermore, as this simple example suggests, we must also possess factual information about the particular X values that will be appropriate to that case, say a year or two into the future. It is no wonder, then, that a theoretical prediction of individual events is exceedingly difficult and becomes possible only at the advanced stages of a science. Witness, for example, the inability of geologists to predict future eruptions of a volcano, such as Mt. St. Helens, that does not have a recent history of similar eruptions. There are simply too many parameters to be estimated and too many unknowns to be supplied.

Contrast this kind of theoretical prediction with that implied by the equation

$$Y_t = Y_{t-1} + u_t$$

or the somewhat more complex equation

$$Y_t = \alpha + \beta_1 Y_{t-1} + \beta_2 Y_{t-2} + u_t$$

In the first equation we merely predict that the value of Y at time t will be the same as that at the earlier time t - 1, apart from a random disturbance term that may throw us off somewhat. In the second equation we use two previous periods with suitable weights β_i to predict to the next event, again allowing for a disturbance term. In both instances we have no theory as to what affects Y, but we rely on the assumption that, whatever Y's causes, they will remain either relatively constant during the interval or, if they are themselves changing, they will continue to do so at a fixed rate. The second of our two equations, since it brings in the earlier period t - 2, would permit us to adjust our prediction to take this time trend into consideration.

Relying on previous values of a dependent variable often works far better in predicting specific future events than does a theoretically based prediction—at least for short-run predictions, in which case it may be reasonable to assume that the unknown causes will remain constant. We use this kind of prediction in our everyday lives when we return to a restaurant that has previously satisfied our taste buds, when we refer to someone as being reliable and therefore trustworthy, or when we use party labels to help us make voting decisions. This is not to say that such atheoretical predictions may not be accompanied by some kind of explanation to bolster our decision, but our primary reliance is on past experience rather than the theory itself.

As long as this type of atheoretical prediction works better than theoretical prediction, why rely on the latter? This indeed poses a strategy dilemma, especially when day-to-day predictions are of extreme importance. Until the theory has been perfected it would be irrational to rely on it. The experts we need are persons who know the detailed facts about the present and recent past, rather than those who merely rely on theory. We Americans, especially, may be biased by our pragmatism in this direction, especially when it costs money and considerable time to arrive at the theory.

Yet, as I have already implied, atheoretical predictions based on a knowledge of the present and recent past are also extremely fallible, especially if longer-term predictions are needed or if an entirely new situation is being faced. This is true because predictions of this nature depend heavily on the presumption that parameters have not changed and that values of independent variables can be estimated well simply by plugging in an implicit constancy assumption that becomes less and less plausible to the degree that one is trying to predict to the more distant future or to different circumstances.

What probably occurs in the case of our hypothetical expert is that the present and recent past are extrapolated into the future but then modified slightly in accord with the analyst's implicit theories of social causation, whatever these may be. Thus the prediction becomes partly a matter of using past levels to predict future ones and partly one of making a crude theoretical prediction. Usually the latter component is not spelled out formally, however, in the sense of stating one's assumptions and theoretical propositions explicitly before applying them to the case at hand.

What we obviously have are several sets of scholarly standards that place priorities at different locations on the knowledge map. Certain social scientists—most notably historians, ethnographers, area specialists, and clinical psychologists—place a premium on detailed, factual knowledge of a relatively small number of cases. For these persons a theory is inadequate unless it can be used to explain each of those cases, almost without error. A statistically based theory that utilizes error terms or probabilistic reasoning is thus not satisfactory, and the effort is placed on explaining the exceptions. Other social scientists are more concerned with establishing statistical "laws" rather than in accounting for specific instances, regardless of how important they may be practically. From this perspective Germany is no more important than South Yemen or Costa Rica. The former group of social scientists tend to be highly perfectionistic with respect to factual details but rather sloppy about what goes into their theoretical interpretations, whereas the reverse set of priorities holds among the latter. Of course, this characterization is an exaggeration. No one wants to be careless about facts or interpretations, but given the extreme difficulties in mastering both approaches simultaneously, something usually gives ground. Our intellectual backgrounds often determine the strongest and weakest points in our armor.

There is perhaps no inherent incompatibility here, merely a set of real hurdles to be cleared. In fact, certain strategies of attack have been suggested from both ends. Although seldom explicitly discussed in contemporary methods texts, what used to be referred to as "analytic induction" is an approach that begins with a small number of cases and detailed facts.[1] In essence, the notion is to attempt a systematic explanation designed to account—in some detail—for patterns in a small number of cases, say nation-states. Having formulated the theory on the basis of these facts, one then looks at another case in some detail. The theory will undoubtedly be found inadequate in some respects, and so the new set of facts is used to reformulate the theory, once more to account for 100% of the variance among the selected cases. Then another case is selected, the theory modified again until it fits the facts, and so on. Presumably, once a certain point has been reached fewer and fewer modifications will become necessary, and one may then become reasonably satisfied with the theory until a new set of facts disconfirms it.

In contrast, we may begin with a statistical analysis and unexplained variance. Let us say that regression analysis has been used to predict a dependent variable Y from a set of explanatory variables X_i. We may now examine the residuals or errors in some detail, perhaps displaying them in a scatterplot in which each case is named. We then try to learn as much as possible about the deviant cases, especially those that lie a long distance above or below the expected values. What do those above it have in common, as distinct from those below? Once we identify certain common characteristics that may be associated with these deviant cases we introduce these as explicit variables into the regression analysis. This will tend to reduce the unexplained variance still further, and we may again display the residuals and try to account for their patterning. After several steps we may again reach a point where the explanatory system works well enough for us to rest satisfied until we find some new data for which it is far less satisfactory.

We might hope to reach the same end point by employing either of these idealized strategies, provided that we were sufficiently patient and insightful to go beyond the first or second step in each instance. Those of us who have studied multivariate analysis are reasonably aware of the many pitfalls and artifacts that can lead one astray by using the second strategy. Unfortunately, the corresponding pitfalls of the analytic induction strategy are not well understood.

One of the most serious problems with the analytic induction approach is in the selection of cases to be studied. In practice this selection is likely to be based on such criteria as the availability of data, the investigator's specialized knowledge, or perhaps some judgments as to appropriateness. One tempting strategy is to select cases on the basis of the dependent variable. For instance, if one wants to understand the causes of peasant revolutions, one may select three settings in which revolutions occurred and three other supposedly comparable ones in which they did not. Such a selection of extreme cases on a dependent variable is known to confound the effects of measured independent variables with those of variables that have been omitted, but this fairly obvious point does not seem to have occurred to historical analysts who have employed this very appealing strategy.[2]

The analytic induction approach might, in principle, be combined with a randomized selection of cases to fruitful advantage. It often runs afoul of the problem of having too few cases relative to the number of explanatory variables, however. As we have noted, one must have considerably more cases than variables before one can also be assured that the high percentage of explained variance is not simply due to this artifact. In effect one is able to take advantage of chance fluctuations in an ex post facto way. If we add another case but also find it necessary to invoke another explanatory variable, we cannot overcome this particular difficulty, though we may conveniently ignore it or assume it away.

Another problem that commonly lurks in the shadows, whether or not we recognize it as such, is that of measurement comparability. The social scientist who prefers to work with a small number of cases and rather detailed information about each may not even perceive the difficulty as being one of measurement, although he or she will almost inevitably be forced to use at least some concepts that refer to similarities or differences among the selected cases. To compare English, French, and Hungarian peasants obviously requires one to make judgments as to exactly which actors should be labeled as "peasants" rather than something else. Similarly, there must be means of distinguishing revolutions from minor protests or other actions deemed to be "nonrevolutions." As one begins to expand the variety of cases selected, say by moving to Africa and Asia, where land tenure systems may be somewhat different, the measurement comparability problems are more likely to become obvious. At that point the analyst may stop with the simple assertion that the settings are so different that comparisons should not be made, but this will almost certainly be a matter of judgment and a question of degree.

This type of problem will be made all the more confusing to the extent that area specialists have been using entirely different concepts and have come to have a vested interest in them. We then encounter debates of the kind that occurred in the field of race relations during the 1940s and 1950s.[3] Is the black-white pattern in the United States really a caste system, or is it something else? Must the notion of "caste" be defined so restrictively that it only applies to India, or are systems castelike to varying degrees? Are there a number of analytically distinct dimensions that may be confounded in India but not elsewhere? Often debates on such matters founder because scholars with vested interests in a certain terminology, or in a particular intellectual domain, rebel at the thought of broadening a concept sufficiently that it may be applied to a variety of settings. In short, the subject of measurement and the assessment of comparability become taboo subjects. Scholars' aspirations regarding factual accuracy may remain high, but their measurement aspirations become extremely low or nonexistent.

Difficulties such as these can be overcome, though we would be ill advised to underestimate their seriousness as well as the human limitations of social scientists who may not really desire to overcome them. One of the first steps seems to be recognizing a series of specific hurdles, including the most tempting kinds of biases characterizing each mode of attack. Recognition that no one approach is entirely adequate, along with more concerted effort to create teams of researchers with complementary skills, would almost certainly help in many instances. We also need to indulge in a careful scrutiny of the various scholarly norms that have influenced social scientists along different strategic paths, norms that have been reinforced by what might best be thought of as folk beliefs among social scientists to the effect that a given path is the *only* one to be followed.

Perhaps most important of all, we need to give far greater attention to problems of conceptualization and measurement, as well as to considerations of specific kinds of problems that arise as one attempts to raise levels of abstraction and to construct theories that tie together findings from diverse fields of application. As we pile up an increasing number of facts in widely disparate fields of knowledge, these tasks may appear so formidable that we cease to make the needed effort. There are numerous signs that we are beginning to approach this state of affairs. At some point in the process, we will cease to reward those few who make the needed efforts and may even tear apart their necessarily primitive products before there has been an opportunity to improve on them. In my view this will be very unfortunate from the standpoint of knowledge accumulation.

NOTES

1. For discussions of analytic induction, see Robinson (1951) and Denzin (1970).
2. In general, design considerations are not handled in macrosociology, perhaps because of the inherent scarcity of cases but also because of the nonrandom availability of data. For a discussion of problems encountered when one selects cases on the basis of dependent variables, see Blalock (1964: chap. 4; 1967).
3. See, for example, Dollard (1937), Myrdal (1944), Drake and Cayton (1945), and Cox (1948).

CHAPTER

5

The Simplification Process: Dilemmas and Strategies

Since social reality is far too complex to be analyzed without introducing numerous simplifications, the obvious question becomes this: What strategies can be employed to carry out such a simplification process so as to minimize various kinds of shortcomings? We may anticipate that any given simplification strategy will have both advantages and disadvantages, as well as adherents and opponents who will be more than eager to extol or condemn one or another such strategy according to their own disciplinary biases, research orientations, or substantive interests, not to mention whatever research or theoretical trends happen to be popular at the time.

In the present chapter I shall briefly examine several simplifying strategies as well as some dilemmas produced in connection with each other. The thesis will be that our primary objective ought to be examining each strategy critically with a view to extending the basic orientations involved in such a way as to bring them closer together rather than being seen as in some sense opposed or counter to one another. I suspect that nearly all of us will agree that this objective is a reasonable one in principle, though we will undoubtedly disagree as to how to proceed in specific terms.

Unless some of the dilemmas encountered are faced squarely, it may be more difficult to achieve a common ground than is often recognized, given the complexity of the real world with which we are dealing and some of the methodological complications already discussed. Therefore we may need to reconcile ourselves to living with many of the ambiguities and dilemmas involved for a considerable time. Certainly these ambiguities and dilemmas will not suddenly disappear if we decide to sweep them under the rug rather than face them squarely. We must also recognize that they are not of our own making but that any possible resolutions will require concerted and prolonged efforts that are unlikely to have immediate and highly visible payoffs. Furthermore, they will have to be faced by entire disciplines, since individual scholars can only hope to deal with them in limited ways.

SIMPLIFIED VERSUS COMPLEX SETTINGS

One choice that practically all social scientists make involves the nature of the phenomena we wish to study. Should we begin by selecting relatively simple settings or social processes, proceeding to more complex ones only after the simpler ones have become reasonably well understood? Or should we plunge into the study of more complex phenomena, though trying to abstract out a relatively small number of analytic ingredients presumed to be of greater significance than others? Should we study aggression in dyads or among hostile nations?

The first of these simplifying strategies involves two possibilities. Either we search for very simple situations as they occur naturally, or we try to construct them in our social laboratories. Instead of studying complex industrial societies, we may begin by examining much simpler ones, as for example hunting and gathering bands that are relatively isolated from other groups. Or we may set up experimental groups of two or three persons, give them simple tasks in controlled settings, and then gradually increase the complexity of the tasks or perhaps the group structure.

Strategies such as these are utilized in some social science research, with a reasonable degree of cumulativeness as long as the basic research operations do not have to be modified in fundamental ways. Thus one may create competitive situations in which rewards are distributed according to a zero sum rule and then look for instances of aggression or attempt to measure group productivity. The size of the group may be increased, with coalitions either being permitted or disallowed, again with a view to studying how aggressiveness or productivity changes. If one has to introduce totally new measures of aggressiveness or productivity, however, or if setting variables must be redefined and measured in different ways, the comparability problem rears its ugly head and generalizability of findings also becomes questionable. If one attempts to compare these simple situations with naturally occurring ones that involve more intense interactions, prolongation of the encounter, or modification of the rewards and punishments, and a gradual alteration of the rules of the game, measurement comparability and generalizability become subject to doubt and theoretical and methodological difficulties become confounded with one another.

Thus the strategy of moving from simple situations to more complex ones, while in principle sound, becomes difficult to employ because of the *simultaneous* introduction of too many complicating factors. The problem is partly one of the investigator's patience and the time lags and expense involved in proceeding in such a slow but systematic fashion. The difficulty is more fundamental than this, however, since it involves measurement and conceptualization problems that, if properly addressed, would require highly complex measurement error models involving a high proportion of unmeasured variables relative to the kinds of indicators that are readily available to us.

The outcome is usually considerable slippage between theoretical construct and one's measures, as well as the likelihood that the *relationship* between construct and indicators is itself a function of the setting. Our measurement assumptions have to vary from one setting to the next, being relatively simple in some but more complex and problematic in others. It is as though the physicist would have to infer mass, temperature, and length on the basis of completely different operations each time the setting changed *and* as though there were no theory or transformation rules specifying how this could be accomplished. The length of the ruler or pendulum would become a variable that could be predicted only imperfectly according to a theory that, itself, was subject to dispute.

What other ways do we have for studying social reality? Instead of trying to locate simple *situations*, we may study more complex ones but impose a series of assumptions so as to *model* these complex situations in simple ways. Thus we develop simple pictures of the complex reality and then try to make definitive predictions from these simplified models or pictures, seeing whether these predictions are approximated by reality. Our problems are now of a different kind since there may be many different models or pictures, all of which yield approximately the same predictions. We must select from a large set of possible variables a smaller number on which to pin our hopes, and we must be very careful about what is going on when we do the selecting so that we do not mislead ourselves.

Such simplifications obviously also are necessary whenever a social scientist attempts a detailed description of ongoing events in a complex social setting. They are necessary especially when historians offer explanations of past events, no matter how carefully they have been described. One is more likely to be liable to the charge of oversimplification, however, whenever a limited number of explanatory variables are explicitly listed or whenever a set of assumptions is laid out for the critic's inspection. When this is done, the usual purpose is to apply some crude form of deductive reasoning based on such assumptions and a finite listing of potential explanatory variables, perhaps supplementing these with a set of empirical data and measures of association and significance tests designed to evaluate the goodness of fit between one's data and a set of predictions based on the assumptions of the model. The simpler such a model, the more vulnerable the argument is to charges of oversimplification, as well as the greater the number of additional variables and alternative explanations that will occur to the hypothetical critic.

An Example: Rational Actor Theories

One of the most perplexing dilemmas faced by nearly all social science disciplines is that of developing simplified models capable of explaining choice behaviors in reasonably complex social settings. How much should one rely on assumptions about cognitive processes to explain such behaviors? And if one attempts to avoid such assumptions altogether, where will this lead? Will definitive predictions be possible, and if so, how will these be derived? Will they be highly dependent on observations of past behavioral sequences and thus basically atheoretical? How can responses to new situations be predicted and explained? If one is to avoid the obviously impossible task of obtaining a complete history of each individual, what kinds of simplifications might be attempted and with what pitfalls? We may briefly indicate some of the issues involved by examining the so-called rational actor approach to explaining such choice behavior.

Approaches such as game theory and many of the maximizing (or minimizing) theories employed by economists are characterized by highly simplistic assumptions about idealized actors. Such actors may have to act in the face of uncertainties, but they are presumed to employ one or another "rational" decision strategy that either maximizes or minimizes some function. One such possibility, for example, is to maximize one's subjective expected utility, $\Sigma p_i U_i$, where the p_i refer to subjective probabilities of certain outcomes, given a particular behavior, and the U_i refer to (subjective) utilities or values attached to the outcomes. But what if actual actors do not behave in such an idealized fashion; suppose they are not rational in this particular sense? One way out is to refer to departures from the investigator's model as being "irrational" or at least as "nonrational." In effect, this defines the actor's behavior to be in error and leaves the matter as such, since, of course, no theory can be perfect. Another alternative is to add complexities to the model. Perhaps not all outcomes delineated by the actor have been anticipated by the investigator. The actor may have a larger set of utilities than the set included by the theory. An actor may, for instance, have a utility for gambling or for varying his or her responses so as to add excitement to the game. If the experiment involves a given financial reward for a particular performance level, perhaps the subject also anticipates praise for one line of action and condemnation for another.

He or she may value some equity principle that involves a reward to one's competitor at the expense of one's own rewards. Is this irrational or a mistake, or simply rational behavior based on a different set of values than those assumed by the investigator?

Suppose the subject miscalculates the probabilities of outcomes occurring as a result of a particular response pattern. Rare events may be overestimated and frequent ones underestimated. Need "subjective" probabilities coincide with "objective" ones as measured by relative frequencies? Perhaps the subject anticipates a change in these frequencies or believes that they are patterned in some way, rather than being a function of some random device. In all these instances we are confronted with the decision of what to call error or what to consider irrational. We must also decide whether or not the errors are sufficiently numerous that the underlying assumptions need to be modified in the direction of increasing complexity. If so, we will need additional information about each actor if our predictions are to remain precise enough to be useful. We must pay a price for the added complexity.

This particular example affords an opportunity to comment on the phenomenon of disciplinary biases or differential proclivities to simplify in specific ways. Economists are notorious for their neglect of "noneconomic" values, which they often lump together and label "nonutilitarian" or as simple "tastes" that are taken as givens. For this they are severely criticized as totally one-sided and naive about human motivation. Perhaps the most reasonable point is that these particular assumptions are highly realistic under a limited set of circumstances, moderately so under others, and almost useless under still others. The trick, of course, is to discover the circumstances for each.

Another way to simplify the rational actor model is to rule out uncertainties or to assume that actors assign subjective probabilities that are close to either zero or unity. Again, this may be a reasonable assumption to make under some conditions but not others. This will undoubtedly be a function of whatever ideologies or belief systems actors have to come to accept, as well as their degree of sophistication in accepting such belief systems only conditionally or under selected circumstances. If I "know" that I will go to hell if I murder a neighbor but to heaven if I kill an enemy, this will make it much easier to predict my behavior using a rational actor model than if it is known that, like Hamlet, I will think deeply about the matter before acting.

Since social scientists differ considerably in terms of the assumptions they are willing to make about the complexity of actors' thought processes, it is no wonder that they also disagree about the advisability of using simple rational actor models. My own view is that such a way of modeling or predicting human behavior is potentially useful, but only provided we can become increasingly sophisticated about actors' utilities and subjective probabilities and can find better ways of measuring them.[1] Many sociologists probably do not share this view, however. The question then becomes this: If we reject such rational actor models in principle as well as in practice, with what do we replace them, and how can we make predictive statements *in advance* of actual behaviors? Is there any alternative, apart from the obvious one of using past behaviors, to predict future ones?

COMBINING SIMPLE GENERIC PROCESSES

As noted, most naturally occurring social processes are highly complex and multidimensional. So are most real-world situations and behavioral choices. Political scientists studying specific issues or political bills encounter problems of comparability because of the obvious fact that issues on which voters or legislators are actually asked to vote are multifaceted. Rarely can such issues be accurately characterized along a simple liberal-conservative or hawk-dove dimension. Legislative bills are designed deliberately to attract as many votes as possible, and this usually means simultaneously appealing to a diversity of interests and priorities. Political candidates obviously also follow similar strategies. Persons socialized in a single religion—Catholicism, for example—are simultaneously exposed to a complex set of experiences, including specific theological doctrines, rituals, individual priests, lay church leaders, peers, and whatever carryover effects these experiences

may have on family members and friends. Some such experiences will be shared with Episcopalians, a different but
overlapping set with Methodists, and still others with all
Protestants. What are we doing when aggregate individuals
according to the simple label "Catholic?" Most certainly,
we are confounding and obscuring a number of specific
experience dimensions that, had they been distinguished,
might have had greater predictive value. What if we lump
Southerners or even white Southerners together? The only
thing that is clear is that the picture is indeed very unclear,
though we may pretend that it is not.

We may often find it desirable to aggregate a number of
different forms of the same behavior, simply because there
are too many variants to be analyzed separately or perhaps
because we believe that some of them may substitute for
others. Thus one does not need to shoot and simultaneously
stab a person; both acts might be labeled forms of aggression
or, if successful, homicide. Yet the manifest behaviors—
here pulling a trigger and thrusting a knife—may be
different. How do we handle such problems? Should we
characterize a mother protecting her child, a young adult
helping an elderly person across the street, and a donor
dropping a five dollar bill into a box as forms of "altruistic"
behavior? We are obviously faced with aggregation decisions
in which some notion of similarity is involved.

Whenever manifest similarities cannot be invoked, one
possibility is to build into the definition of the generic
concept—here a behavior—some assumptions about a set
of causes (e.g., a motivational state) or the supposed effects
of the phenomenon in question. Thus one may define
aggression as behavior the intent of which is to injure
another party and altruism as behavior intended to benefit
someone else but not oneself. Or these behaviors might be
defined in terms of their usual effects: aggression as
behavior usually resulting in injury and altruism as behavior

usually benefiting someone other than oneself. A causal theory is thus invoked in order to classify the behavior. This theory may be more or less simple and more or less realistic and open to challenge. Unfortunately, "pure" forms of most general behaviors can seldom be found in complex situations. Many acts of aggression are also instrumental in achieving other ends, even including the benefit of the target actor. Apparent altruism usually also benefits the self in terms of guilt reduction or a sense of satisfaction in having helped someone else. Thus two observers, each witnessing the same act, may classify the behavior in different ways according to the assumptions made in their causal theories. A cynic may tend to see many more acts of aggression and many fewer of altruism than someone who assumes the best about human motivations. Their aggregating assumptions and decisions may then be quite different. Unfortunately, there may be no empirical way of deciding which is more correct. This is but one instance of the general point that measurement decisions almost always involve theoretical assumptions, whether explicitly recognized or not. It also points to the great difficulties encountered in dealing with complex real-world phenomena that are multidimensional in nature.

One theoretical alternative is to attempt to specify a number of "pure" processes, in which simple assumptions about motivations or consequences can be made, and then to test these theories by either locating extreme examples in which these assumptions seem plausible or by constructing laboratory situations characterized by such simplicity.

Thus economists theorize about perfect markets involving large numbers of independent actors who are not permitted to form coalitions or conspiracies and who have simple motives. They are not supposed to favor friends or relatives, or be coerced in their dealings with others, or make complex assumptions about others' future behaviors. Under such idealized conditions it then becomes possible to make certain assumptions about maximizing behaviors and to follow through the implications to make falsifiable predictions. If such predictions do not hold in practice—and they usually do not—at least one then knows that something is wrong. Perhaps the actors are not acting "rationally." Perhaps, however, they *are* reacting rationally by secretly violating the rules of the game, say by making illegal bargains and bribing potential agents of control. Or they may be coercing one set of actors by paying another set, and so forth. The advantage of simple laboratory experiments, of course, is that many (though not all) of these potential complications can be ruled out or at least accurately observed.

A wide variety of rather general social processes have been identified and examined in their purified forms. Market behavior is, of course, a special form of competitive behavior among similar actors, each desiring to maximize his or her share of a scarce and valued good, subject to certain constraints. It is also a special form of social exchange behavior, but there are other exchanges that involve different types of rewards and punishments, such as prestige, sexual favors, friendship, or protection. Conflict, itself, may be defined as involving the mutual exchange of negative sanctions, the most extreme form of which may involve death. There are many of forms of interaction that may best be analyzed in other terms—for instance, power and dominance, socialization or learning, coordination, cooperation, approach and withdrawal, or the allocation of resources.

One tactic that may be followed is that of developing theories about each of these pure processes and then testing them by locating or creating simple situations in which the other processes may safely be neglected. One evaluates game theory propositions by letting actors play games, but without permitting them to introduce guns to threaten opponents or to pay off referees or judges to induce them either to change the rules in midstream or to neglect deviations of certain types. If part of the game is to bargain with potential coalition partners, this possibility may be built into the situation; but one would not, for instance, encourage actors to make bargains involving sexual favors as supplements to money payments. Nor would one permit fistfights among participants or allow them to withdraw from the game prior to its completion.

All of these extreme examples make a point: Naturally occurring social processes involve *combinations* of simpler ones. Furthermore, the complications in each specific situation are usually unique, at least in the sense that it would require a large number of replications to encounter the same combination of circumstances more than a few times. Where instances are not sufficiently numerous to permit statistical controls, it may prove almost impossible to unconfound the separate processes. Power considerations usually intrude on market processes and vice versa. We rarely find other situations comparable to the peculiar ethnic-religious-economic-territorial disputes that characterize the current conflicts in the Near East. From the standpoint of world stability this is indeed fortunate, but from the standpoint of using such situations to understand conflict processes it is not.

A rather obvious strategy for extending our theoretical analyses of simple processes is to examine them in pairs and then triads, but to my knowledge this has been done only incidentally and sporadically; it has not become a consciously developed research and theory-building strategy.[2] Thus one could link dominance and coalition theories with those attempting to account for learning or the socialization of new members into a group. Exchange theories and theories of deviance might be joined to see what complications in each approach are encountered. Conflict and power processes are often studied together, although they are analytically distinct. Sometimes conflict theories are conjoined with learning theories but not in any systematic way. Approach and withdrawal processes are obviously linked to both exchanges and conflicts, but again we do not seem to have extracted the implications of such dual processes over and above those of each one separately. Norm or rule development processes are sometimes loosely linked to power processes but not to approach-withdrawal ones.

Since naturally occurring processes often involve as many as four or five simpler processes simultaneously, once we have obtained a sufficient understanding of the complications introduced by the joint operation of particular pairs of processes, we could then begin to introduce a third and perhaps even a fourth. Coalition formation, for example, is obviously linked to power considerations, exchanges, and normative regulations. Soon we would encounter a familiar phenomenon, however: too many complications at once and not enough cases or replications to go around. We would also find, undoubtedly, that several phenomena were confounded in a large number of instances, so much so that social scientists who had confined their attention to these cases alone may never have recognized that they were analytically distinct.

A separate problem is how actors *perceive* these social processes and how they go about simplifying their working theories about them. Such perceptions are undoubtedly affected not only by vested interests but also by ideological systems that encourage specific kinds of simplifications. Thus Americans may tend to perceive market exchange processes as independent of socialization processes or power considerations, whereas citizens of the USSR may see them in an entirely different light. Residents of prosperous, developed countries may perceive guerrilla-type social movements in terms of illegitimate uses of violence to extract economic and political concessions, whereas residents of the local area may perceive them as responses to economic exploitation on the part of powerful governments supported by economic elites.

Social scientists must take these simplified belief systems into account in explaining differential responses to complex processes, but this does not imply that our analyses need to involve the same simplifications made by involved actors. Of course, to the extent that we are unaware of our own biases and simplifying assumptions we may be misled into making incorrect analyses of these phenomena. Thus one must distinguish between theories about the assumptions actors are making on the one hand and those required for an adequate interpretation on the other. The latter, when made explicit, may also help us attain a better understanding of the former. We certainly need a better understanding of the processes through which human actors make decisions in the face of complexities and unknowns, and how whatever biases they may have as a result of their socialization and social positions affect these decision processes.

In all of this we see the need for conceptual clarifications and the explicit statement of whatever assumptions we are aware of making. For instance, there is often confusion over terminology that suggests that certain phenomena are polar opposites, merely because they are empirically related in a negative fashion. Thus conflict and cooperation, competition and cooperation, socialization and coercion, or exchange and withdrawal may sometimes be thought to be extremes merely because instances of their joint occurrence are rare. Worse still, perhaps, the concepts may not have been defined with sufficient clarity for a reader to determine whether or not an author is using some terms interchangeably (here as opposites), or whether an empirical relationship is being presumed, or even both! Such a state of confusion is especially likely to occur whenever one is dealing with ideologically "loaded" words such as "exploitation," "coercion," or "domination."

SOME METHODOLOGICAL ISSUES

It has been emphasized that theoretical and methodological issues are often intertwined. In the remainder of the chapter I will discuss three issues that might be considered primarily methodological but that all have implications for theoretical questions. All directly involve dilemmas of simplification, which we have stressed will always be necessary before definitive predictions can be made. The trick is to avoid premature simplifications, ideally by allowing one's data to permit tests of simplifying assumptions before they are imposed on the analysis. Sometimes data may be collected to permit consistency checks on such simplifying assumptions, and indeed the methodological literature on scaling techniques deals specifically with such procedures.[3] But we also often find ourselves in the position where, lacking such supplementary data, these assumptions must be made a priori. It then becomes tempting either to hide them from the reader's view or to pretend to ourselves that assumptions we have had to make for reasons of expediency are also justified on theoretical or substantive grounds.

I shall first consider the problem of errors that arise in scaling applications in which the assessment of dimensionality is the primary objective. I then turn briefly to the complex problem of aggregation and the usually implicit homogeneity assumptions this requires. Finally, I conclude the chapter by raising the more general methodological dilemma of the tradeoffs that almost always exist among the objectives of constructing theories that are simultaneously parsimonious, precise, and rejectable, and also sufficiently realistic to explain real-world phenomena of general interest.

Simplifying Assumptions and "Errors"

Some 30 years ago Clyde Coombs (1953) pointed to what he referred to as a fundamental problem in connection with scaling theory—the problem of what to call "error." He argued that the matter can be posed by this question: Do we know what we want, or do we want to know? If we are willing to *assume*, say, that a given set of items can be placed along a single continuum representing levels of difficulty or perhaps degree of favorableness toward some object, then actors may be expected to respond in highly predictable ways to the items. If, for example, a series of math questions can be ranked from easiest to most difficult, then someone who misses two questions will necessarily miss only the two most difficult ones; someone missing three will miss these two plus the third most difficult one, and so forth. If the items represent a "social distance" scale indicating willingness to associate with a member of a different ethnic group under circumstances involving increasing levels of intimacy, then we expect that someone willing to marry a member of that group will also be willing to live next door to one and to work with one; someone who is unwilling to intermarry but willing to live next door will be willing to work with one, and so forth.

If a hypothetical respondent misses an easy math question but gets a more difficult one, or if he or she is willing to intermarry and work with a person of another ethnic group but not to have one next door, then something peculiar is taking place. We count such responses as errors, at least if they are relatively rare. But if a large number of such deviant patterns emerge, we may begin to question our underlying assumption of a single continuum of difficulty or social distance. Perhaps the math questions tap two different kinds of skills, so that some questions are more easy for one person than another, whereas the latter individual has an easier time with a different set of items. Perhaps there is more than one dimension along which social preferences may be ordered, so that the "deviant" individual was responding in an entirely "rational" fashion.

Would we then want to call the responses "errors"? Or would we want to make our underlying model more complex, say by adding a second or a third dimension? In doing so, we would then have to make our predictions more complex and also less precise. We would also need to collect additional information about our respondents (or test takers) in order to predict their behaviors more precisely. Clearly, we give up something when we increase the level of complexity of our assumptions, and so we need well-defined criteria for making such decisions. Fortunately, in some technical bodies of literature—such as scaling theory—such criteria indeed exist. As one might expect, we usually discover that multiple criteria become necessary and that such criteria often force one to make difficult choices, none of which is entirely satisfactory. When the methodological principles are understood only vaguely, however, added complexities usually generate so much theoretical confusion that we begin to tolerate errors as a matter of course, rather than viewing them as starting points for a more systematic analysis.

Similarity Judgments and Aggregation

Whenever we compare individuals for statistical purposes, we make a series of implicit or explicit assumptions about them. In particular, we make homogeneity or similarity assumptions to the effect that not only are these individuals similar with respect to certain designated characteristics such as age, sex, or race, but they also *respond* similarly to relevant stimuli. Consider the simple regression equation

$$Y = \alpha + \beta X + u$$

where u represents a disturbance term assumed uncorrelated with X. Giving a causal interpretation to the equation, we assume that a unit change in X produces an average change of beta units in Y. Such an interpretation makes sense only if we are willing to assume approximately the same beta values for all individuals under consideration, however. A more correct specification of the model might be

$$Y_i = \alpha_i + \beta_i X_i + u_i$$

where the subscript i refers to the ith individual, and where we are allowing for a possibly different value of α_i and β_i for each unique individual. In order to estimate the separate coefficients for each person we would need repeated measurements for each individual, information we do not have available in most instances. We therefore make the assumption that α_i and β_i are the same constants for some set of actors and we pool the information over such individuals to obtain our estimates. A critic challenging such an assumption could, if the data were made available, disaggregate and distinguish among subsets to obtain possibly different estimates of the parameters. For instance, individuals might be separated by race and sex to see whether the betas differed across these types of individuals.

Clearly, however, *some* individuals would need to be pooled and questions could similarly be raised about the legitimacy of the implicit simplifying assumptions being made. Perhaps not all black females respond to changes in

X in the same way, in which case further subdivisions might be made. At some point one would have to stop and make the simplifying assumption that those subsets of individuals who were ultimately pooled were homogeneous with respect to the relevant coefficients. Given the fact that in multiple regression equations there will be several different independent variables, with the obvious possibility that subsets of individuals homogeneous with respect to one coefficient may not be so with respect to another, we can easily anticipate a number of complications that could arise.

Often the raw data necessary for aggregation are not available and the aggregaton has already been carried out before the data have been presented to us. Suppose, for instance, that individuals have been grouped together according to the census tract, county, or even the state in which they happen to reside. Even if we have separate data for males and females and by race, age, or income level, what assurances do we have that individuals grouped together by proximity will be similar with respect to their responses to changes in some independent variable that probably has not even been considered by those making the aggregating decision?

Why are people located in space the way they are? Perhaps they have moved into the area—especially if it is a small one such as a block or a census tract—as a *result* of some of the variables being considered in the explanatory model or even the dependent variable being explained. We will have aggregated in terms of a variable—spatial location—that belongs somewhere in the theory, but we know not where! Perhaps some people moved into the area whereas others did not, in which case we will have aggregated individuals with very different life histories that may be relevant to the phenomena we are trying to explain.

Aggregation issues such as these are highly complex and technical, but the implications are clear.[4] Unless one understands something about which individuals have been grouped together, for whatever practical reasons, there will be unknowns that are not at all likely to cancel each other out. In fact, it can be shown that under plausible assumptions about such aggregating criteria we are likely to *amplify* any biases that may have existed prior to aggregation, but we will not be able to discover this fact unless we are in a position to compare the aggregated data with data based on individual-level information.[5] If we have only the former but if our theoretical interests are centered on the individual-level phenomena, we are likely to be misled without ever recognizing this to be the case.

Once more it seems advisable to comment on some natural temptations. If the necessary information is unavailable to us—in this case, the disaggregated data—the temptation is to do away with the problem by pretending it does not exist or by making some unrealistic assumptions that we hope our readers will accept uncritically. Chances are, they will be more likely to do so if they are unaware of the methodological problem in the first place or if they are distracted by lengthy discussions of specific measures, sources of information and possible biases in them, or a highly complex and apparently sophisticated data analysis that simply neglects to mention this particular kind of problem.

The intellectually honest posture, of course, is to admit to the difficulty, try to suggest at least the direction of possible biases being introduced, and suggest specific alternative ways of proceeding in instances where better data might be made available. This means that we are doing the job the critic is supposed to be performing, however. Unless this becomes a highly salient scientific norm among social scientists, we may expect that it will be honored more in the breach than in practice.

The Parsimony-Testability-Realism Dilemma

Social movements often fail because their objectives are too ambitious or unrealistic to be achieved with the level of resources or commitment that is available within a reasonably short time span. They also fail because their objectives are multiple and incompatible to varying degrees, so that disagreements as to priorities may immobilize or confuse the rank-and-file member. Much the same phenomena can occur within an academic discipline confronted by a number of major hurdles that cannot be overcome simultaneously or that force one to make a series of choices that inevitably dissatisfy major factions. Whenever we propose desirable scientific goals that involve fundamental incompatibilities given whatever resources and knowledge are currently available, we run the risk of encouraging many kinds of intellectual disputes.

Most of us would probably agree that, ideally, it is desirable to propose theories that are at the same time relatively simple, capable of being rejected whenever they are not compatible with one's data, and realistic in accounting for reasonably general phenomena of social importance. Yet we may not be fully prepared to cope with the consequences if one of these criteria is incompatible with the others. In particular, a sacrifice of any of them may bring about immediate, overly critical reactions from those who do not like the conclusions reached, the research methods used, or the assumptions that inevitably must be made. Yet these same critics will not be able to produce alternative formulations that, themselves, are not subject to countercriticisms. We may then be treated to interminable debates that never seem to confront the fundamental dilemmas involved. If we are ever to achieve a working consensus on the objective of reducing self-destructive debates and unreasonable criticisms, it seems essential that such dilemmas be squarely faced.

Suppose we begin with a reasonably simple formulation involving a single phenomenon or dependent variable to be explained by a set of explanatory factors or independent variables. Whether we represent such a theory by a series of propositions or an equation system, the theory must then be "closed" if one is to make specific predictions or to estimate the relative impacts of the explanatory variables. For instance, in the case of recursive systems we look at a series of dependent variables one at a time, in succession, beginning with those we take to be most causally prior. Even at the outset, however, the argument can be criticized as incomplete, in the sense that it cannot handle any explanatory factors taken to be prior to the ones we have included.

Further assumptions must be made if the separate equations or verbal arguments are to be treated individually in this sequential way rather than simultaneously. It is all well and good to make seemingly profound but vague claims to the effect that "the whole is greater than the sum of its parts" or that the factors are "inextricably bound," but unfortunately there is no place one can go with such assertions in the sense of making falsifiable predictions. So right away we see that falsifiability requires a sufficiently restrictive set of untested assumptions that many critics will be in a position to reject the argument out of hand as being overly simplistic. Yet the complications they may wish to introduce may result in underidentified situations in which estimation becomes hopeless, or they may force the introduction of other simplifications we ourselves may not wish to accept.

If, for instance, one wishes to allow for almost unrestricted interdependencies in which feedbacks occur more or less continuously rather than at determinate and discrete intervals, then the effects of the resulting set of mutually interdependent variables can be disentangled only by making further restrictive assumptions, either about lag periods or about certain exogenous variables that are assumed *not* to be affected by any of the endogenous.[6] Furthermore, one must then be in a position to measure each of these exogenous or lagged endogenous variables. Thus the more one allows for complexities among the mutually interrelated endogenous variables, the more one must rely on restrictive assumptions about the causal connections of exogenous variables to these endogenous variables. Attention must also be paid to measuring these additional variables with minimal error. Otherwise, the presence of substantial measurement errors will introduce a sufficient number of new unknowns to produce an empirically hopeless situation.

What all of this implies, of course, is the need to compromise between the desirable objectives of keeping a theory simple enough to be testable and yet complex and realistic enough that the assumptions required for testing and estimation are also sufficiently reasonable that critics will not simply laugh at one's naivete. What also must be recognized is that many such required assumptions will be either inherently untestable or at least untestable with the data at hand. Such assumptions must be agreed on in order to achieve consensus on any estimates we make or the causal model(s) that are most consistent with the data.

Unfortunately, there is an inverse relationship between one's stubbornness or unwillingness to make such assumptions and the number and precision of rejectable predictions one can make. For example, the more willing one is to make zero assumptions to the effect that some variables do *not* directly affect others, the larger the number of null hypotheses that may be tested. One's aim must be to make a sufficient number of restrictive assumptions to permit enough rejectable predictions that one's critics will be reasonably satisfied. Yet if one makes too many such untested assumptions, the theory may be rejected as being far too simplistic.

Given the virtual consensus among social scientists that reality is indeed highly complex, it seems strange that the principle of parsimony is also so highly appealing. Clearly, complex theories are both difficult to formulate and test and taxing of our mental capacities. Insofar as they also contain ambiguities or loopholes, they are also more difficult to reject than are simpler alternatives. This is particularly the case for verbally formulated theories that consist of lengthy arguments containing numerous undefined terms, "weasel words," and vaguely formulated qualifications.

Data analyses of a complex nature usually require an understanding of relatively difficult statistical procedures, with the result that some will look on them uncritically with awe whereas others will reject them as outlandishly over-complicated. Thus sometimes our vested interests work in the direction of complexly stated flexible theories but simplistic data analyses that cannot possibly do justice to these theories. My own position is that if we believe that social reality is indeed complex, then both our theories *and* our data analysis also need to be reasonably complex, though not so much as that the estimation situation becomes intractable. This, in turn, implies the need for more complete data sets than are usually available or that can be obtained with limited resources. The dilemmas involved are thus very real; we must pay a certain price every time we raise our scientific aspiration levels.

NOTES

1. This point of view is elaborated in considerable detail in Blalock and Wilken (1979). Unfortunately, the measurement of both subjective probabilities and utilities is sufficiently difficult, even in carefully controlled laboratory experiments, that there has been considerable reluctance to apply rational actor theories in more realistic complex situations involving multiple goals and multidimensional outcomes.

2. An important exception to this generalization is the work of Blau (1964), Emerson (1972), and others linking the literature on power and dependence to that on social exchange. It is precisely this type of theoretical work, in my opinion, that needs to be encouraged.

3. For an insightful discussion of many of the fundamental issues involved in this type of measurement area, see Coombs (1964).

4. For detailed but more technical discussions of this general problem, see Hannan (1971) and Langbein and Lichtman (1978).

5. See Hannan and Burstein (1974), Langbein and Lichtman (1978), and Hammond (1973).

6. These facts about simultaneous equation systems are now fairly well diffused throughout the methodological literatures in political science and sociology, as well as economics, but their *implications* have not yet been incorporated into most actual data analyses in the former two fields and, to my knowledge, in the remaining social sciences. For discussions, see Christ (1966), Golberger (1964), Johnston (1972), Duncan (1975), Namboodiri et al. (1975), and Hanushek and Jackson (1977).

CHAPTER
6

Some Defects in Our Intellectual Culture

This chapter is concerned with what I consider some negative characteristics of sociology that I assume apply to the other social sciences as well, though perhaps to varying degrees. Many positive accomplishments have been achieved in the past several decades, but my aim is not to deal with them. This will obviously imbue the discussion with a tone that is far too harsh a condemnation of current practices. Yet, as insulated scholars, we often become so accustomed to whatever practices we see around us that we may fail to recognize that many of them are dysfunctional to the development of scholarship, as well as to the images of the social sciences that we portray to others, including our students.

Scholarly norms will vary from discipline to discipline as well as from time to time within a single discipline. Given this fact, it becomes important to examine those practices that we believe to be both correctable and harmful to the accumulation of knowledge. It is in this spirit that I shall attempt to characterize certain features of our academic heritage that seem to me especially noteworthy and closely bound to at least some of the difficulties with which we are currently faced. In the final chapter I shall attempt to end on a more positive note.

DISGUISING OUR WEAKNESSES

Given the obvious rewards for contributing to knowledge— or appearing to do so—and for gaining credibility among policymakers and the general public, it would be strange indeed if social scientists did not develop vested interests in seeming to be in possession of vast stores of knowledge, or at least knowledge that is far superior to that of laypersons. As Robert Merton (1968) argued several decades ago, whenever ends become highly important whereas means come to be evaluated primarily in terms of their technical efficiency, one may expect strains toward deviance from whichever legitimate norms are less efficient than illegitimate ones. There also will be strains to modify the rules of the game and to redefine the norms of legitimacy so as to favor whatever means are most efficient.

The pressures toward deviance are considerable, particularly where social scientists are attempting to follow the science model and where we find ourselves being compared with persons in fields that have had a head start in terms of cumulative knowledge building. Laypersons tend to look at science in terms of *results* rather than as a *process* of learning. They want to see applications with immediate payoffs, especially when they are paying the bill. In a highly competitive world of grants and contracts, and particularly during periods of relative decline in research funding, there is an almost automatic advantage to those who can promise the most and who can convince an audience that they have the technical abilities to "produce." Furthermore, we social scientists will find it to our short-term advantage to modify our own scholarly norms so as to permit us to engage in the competition with relatively little guilt or feeling of intellectual insecurity. How can this be accomplished?

The devices I shall mention are all obvious ones, but this does not mean they are consciously or cynically employed. Indeed, a social science is successful if it can convince its own members—and especially its graduate students—that the practice concerned is not only legitimate from a scholarly standpoint but also professionally necessary and worthy of being defended from outside attack. The reader will recognize in my own list of such devices any number that have been held up to ridicule, especially by laypersons. Yet is must also be kept in mind that each of these practices, if not carried too far, is either necessary from the scientific point of view or desirable in terms of communication within the profession.

Use of Jargon and Ideological Language

Since it is difficult for outsiders to distinguish between technical terms that are useful in a scientific context and clearly understood by members of a profession and words that are bandied about in a field but that have no clear meaning, the use of "big words" can often substitute for knowledge. Consider, for example, the differences between statistical terms such as covariance, homoscedasticity, and multicollinearity—all of which have well-understood technical meanings—and sociological concepts such as functional integration, institution, or cultural pluralism—which do not. It often requires considerable effort to master the vocabulary that characterizes a particular school of thought, such as ethnomethodology or Marxist sociology, or a given social theorist, such as Talcott Parsons. Once the terminology has been adopted, even where it has led to considerable conceptual confusion, there will be strong interests favoring its retention. Since others will favor a different set of concepts, however, we find that adherents of the numerous schools of thought will tend either to ignore one another or to take "potshots" at each other's terminology without ever making a serious effort to achieve mutual understanding.

Outsiders may fail to recognize the ambiguities and lack of consensus on terminology simply because they cannot take the time to discover these ambiguities for themselves. As insiders we become accustomed to the confusion, lowering our aspiration levels in terms of our search for clarity and consensus on terminology. Indeed, any thrust toward achieving greater consensus on terminology may become interpreted as an attempt to impose orthodoxy. A state of terminological confusion becomes a normal state of affairs, and incoming members of the profession are socialized to accept it.

When combined with ex post facto interpretations of past events, slippery concepts afford an excellent opportunity to provide learned interpretations that are completely unfalsifiable. Concepts that shift their meanings or that can be conveniently interpreted in many different ways may be used to provide analyses that seem to go far beyond the commonsense level and, therefore, that are convincing to the general public. Freudian psychology has been notorious in this respect. Since there can be reaction formations on top of reaction formations, since meanings of events can be distorted in a wide variety of ways, and since motives can be hidden deep in the subconscious and actions and reactions almost indefinitely delayed, we have in Freudian psychology almost the ideal form of a theory for explaining virtually everything, especially if the explanation is sprinkled with terminology that the layperson presumes can be understood by experts. If Freudians were the only ones guilty of this kind of practice, the social science community could rest reassured. Clearly they are not.

A closely related device more characteristic of the literature directed *within* a profession is that of ideological intimidation, although perhaps this phrase is a bit too strong to apply to some of the relatively more subtle processes at work. In sociology, especially during the 1960s, our liberal sensitivities were aroused by even the slightest hint that the whites among us were racists, the men were male chauvinists, or non-Marxists were tools of the establishment or neoconservatives. One way to inhibit a serious scholarly exchange— say about a particular book—was to suggest that someone else's comments were obviously racist in nature, or that they involved instances of male chauvinism or a conservative political philosophy.

It was not so much jargon that was at issue, although this was involved too. The credibility of scholars with opposing viewpoints was placed on the line. Their supposed biases, rather than genuine doubts about ideas, were claimed to be at stake, and with it their credibility as scholars. It became easier simply to ignore a line of work rather than to criticize it constructively. Such forms of intellectual intimidation also occur in more subtle forms. This has been the charge laid at the doorsteps of so-called establishment social scientists, and probably there is at least some truth to it. Ideological slogans facilitate such processes, especially during periods of intellectual crisis within a discipline.

Providing Endless Factual Information

Another device that is sometimes encouraged by disciplinary norms is that of claiming that an extremely large number of facts must be mastered before an adequate understanding can be achieved. A reader is presented with so many statistical tables or such a detailed description that exhaustion sets in before the most significant facts can be interpreted. In effect, a surfeit of facts is made to compensate for the paucity of ideas. As a result, nearly all factual information is counted equally. The reader may also be supplied with copious footnotes that cite virtually every study that has been done on the subject, regardless of its relevance or quality. It is almost as though the reader is being challenged to stick it out long enough to await the conclusions. There is also a ploy here, since anyone with insufficient patience to absorb all such factual material is presumed to be unqualified to judge the final product. Historians and ethnographers seem to have developed norms that especially encourage this type of scholarly reporting, but certain quantitative fields, such as demography, have similar qualities. They are also often characterized by other social scientists as being fields that are strong on facts but weak on theory.

Relying on Overly Technical Reports and Analyses

Technical analyses are obviously necessary in many lines of inquiry. This implies that reporting the results of such analyses to relatively unsophisticated audiences will be a difficult task at best. I am referring, however, to the device of overwhelming one's audience with technical expertise that is really not needed for the problem at hand or that is designed to distract the reader's attention from the lack of substantive findings or theoretical interpretations. In one sense, this device is similar in impact to the one previously discussed, in that the reader or client is being primarily impressed by an intellectual tour de force that amounts to overkill and that serves primarily to deflate the reader at the expense of the expert. In effect, the reader begins to assume that because the expert possesses skills that *may* be relevant to the questions at hand, the expert's judgment and conclusions should not be challenged, except by other experts. The intellectual leap from technical analysis to conclusions may be a considerable one, but the reader will not be equipped to understand what has taken place.

Ignoring Missing Information

The expert or technical specialist will be in a much more favorable position than the outsider to know what information has *not* been provided, what variables are either missing or poorly measured, how the selection of observations has been made, how the relative importance of variables has been assessed, and many other potentially important considerations that have entered into the study or report. Ideally there is an obligation to report such matters in some detail and to assess their implications. Where a report is already too long, however, or a writeup more technical than the reader can follow, the exhausted reader is unlikely to pay careful attention to further details, disclaimers, or technical appendices.

In a sense, the social scientist may be intellectually honest in mentioning problems and shortcomings while relying on readers' impatience and intellectual laziness to result in their actual neglect. In statistics, for example, many neglected variables are hidden in error terms that appear in regression equations or as conditional probabilities in contingency tables. Unless they are really emphasized and discussed early in a report, one may safely assume they will be ignored, except by those specialists who disagree with the conclusions reached. Where disciplinary or other biases exist regarding entire lines of inquiry or sets of explanatory factors, such missing information may seldom be noticed unless forecasts made by the social scientists are consistently off the mark. Social scientists in other fields frequently note that economists are often guilty of this type of practice—namely, leaving out all sorts of factors merely because they are either difficult to measure or not encompassed by economic theory.

Ignoring Simplifying Assumptions

This device is basically similar to the prior one, but it is nevertheless worthwhile to distinguish the two. Generally speaking, the more information that is missing, the more untested assumptions we have to make in order to compensate. Of course, *any* explanatory system must necessarily be based on a large number of assumptions, only some of which can be justified in terms of existing data. Social scientists differ in terms of the degree to which their assumptions are placed "out front" for the reader to challenge, however, and none of us can be fully aware of all the assumptions we are making.

This being the case, whenever one is in doubt about an assumption, the temptation is to hide it from view, possibly by using vague language or simply playing it down by embedding it in a number of innocuous assumptions or a technical discussion that most readers are unlikely to follow. Some notion of "robustness" or the seriousness of departures from assumptions becomes crucial here. One's analysis may be relatively insensitive to major departures from one set of assumptions and yet highly sensitive to small departures from another, but if the reader does not know this, the net result may be that he or she will treat all such assumptions as having equal weight.

One of the strong points of mathematical and statistical modeling, as compared with loose-knit verbal accounts, is that scholarly norms in mathematics and statistics virtually require that assumptions be made explicit. In mathematics they take the form of axioms that play essential roles in the formulations. In statistics—as, for example, multivariate analysis—*certain* assumptions are made explicit whereas others may tend to be neglected. For instance, the assumption of no measurement errors is virtually always made in multivariate analyses unless explicit measurement error models are invoked. But the statistician, as an outsider, will commonly take the investigator's word that measurement decisions are not at issue, so that, for example, a classification scheme will be taken at face value.

A random or other type of probability sample is presumed in order to carry out significance tests, but neither the statistician, as a consultant, nor a reader of the report will know the degree to which the ideal sampling design has been approximated in practice. A trained reader may ask a series of questions about how many call-back interviews were attempted, what percentage of respondents refused to answer, and how ambiguous responses were coded. An investigator who fails to report on such matters can often get by without such questions ever being asked, however. Many times naive sponsors of contract research will make awards to the lowest bidder without even realizing that many corners will be cut in conducting research, especially when no one asks the right questions about design and cost considerations.

Judicious Selection of Cases or Facts

The specialist or expert is in an excellent position to mislead naive readers through the initial selection of cases to be studied or facts to be reported.

This temptation will be especially pronounced whenever the social scientist has a theoretical axe to grind or whenever a client or agency wishes to reach a certain conclusion, regardless of the facts. The danger here is almost the opposite of that of overwhelming a reader with endless facts or technical arguments. The investigator selects for them and then draws the inevitable conclusions. Sometimes the exact reasons for the selection will remain obscure, even in the social scientist's mind. For instance, suppose one wants to compare five or six societies. It would obviously be impossible to gain expert knowledge about 50 or 60 and then use a random device to select among them. Usually we obtain some in-depth knowledge of a very few societies and then select several others for contrast.

Although the issues are technical, one major fallacy in this regard is to select on a *dependent* variable with a view to explaining why differences have occurred. For instance, if one selects three societies that have undergone major social revolutions and three others that have not, it is reasonable to use this design to study the *consequences* of these revolutions or nonrevolutions. If one wants to study their *causes*, however, such a design—which involves the selection of extreme cases on the dependent variable—will lead one to confound all sorts of causes, including some that have not been noted by the investigator.[1] Whether this is a consciously employed device or an unnoticed artifact, it will have the same misleading effect. Only experts will be in a position to detect the problem, however, Nor will laypersons have much of a basis for deciding whether or not omitted cases would have yielded different conclusions.

Making Only Ex Post Facto Interpretations

Given the multivariate nature of social reality and the expert's ability to select out explanatory factors that are *consistent with* some specific event that has already occurred, it is relatively easy for the specialist to look good if only post factum explanations are required. Why did the Roman Empire fall? There are any number of plausible reasons, along with many others that are not. Why did Hitler rise so rapidly in Germany, and why did he turn to the Jews as scapegoats? Why did Johnny go astray and Mary succeed so well? And why didn't Blacks riot in the streets in 1982 whereas they did so in the 1960s? Answers can always be provided, not only by reputable social scientists but by popularizers and demagogues as well.

The expert possesses a special advantage in that he or she can usually invoke some little-known fact about the situation that, when added to a few plausible assumptions, can be made to seem like a profound explanation. The question is, or course, could the phenomenon have been predicted *in advance* with the same factual information and set of assumptions? Usually not. Knowing this, we hedge our bets, sometimes with relatively shocking predictions that are too imprecise to be verifiable: "There will be a bloodbath, somewhere, sometime." "Resentment is building up but has not yet reached the surface." "There will be 'ripple effects' of this policy, but they may take many forms.''

It should not be thought that these and other devices involve conscious plots to deceive the public, or even ourselves. As noted, many of them entail simple exaggerations of perfectly legitimate scientific practices. We *do* need specialized technical vocabularies that others may refer to as jargon. We need to present enough facts to justify and qualify arguments, and our analyses and writeups often have to be technical. Providing such details involves the risk of tiring the lazy reader, as does stating assumptions and discussing the nature of missing information. Every discipline requires scholarly norms to assure the integrity of research, and compromises with sponsoring agencies and one's readers must inevitably be made.

What is most important in considering these temptations is that we become aware of their potential harmful and misleading effects, as well as the temptation to hide our defects from view and to oversell ourselves in subtle ways that will, in the long run, work to our own disadvantage. A major step toward scholarly maturity comes when one recognizes potential hazards and takes constructive steps toward their elimination. Appraising carefully and honestly what is going on is one such step. The second, and perhaps most difficult, one is to find effective ways of combating each specific pattern.

THE NONELIMINATION OF
INADEQUATE THEORIES

One of the basic premises of scientific research is that one proceeds by eliminating or modifying those theories that fail to make correct predictions. Since several alternative theories may yield the same predictions, one can never really be assured of a theory's adequacy except in the negative sense that it has successfully resisted elimination in spite of its ability to state falsifiable predictions. What if theories cannot be eliminated because they either make very few falsifiable predictions or because, when predictions are apparently false, one can easily salvage the theories by pointing to inadequacies in the research—by implying that the proper conditions under which the theories hold true have not been met, or by invoking lists of "other variables" that have not been controlled?

First of all, there are a host of reasons why empirical tests of most theories are inadequate. The multivariate nature of social causation assures us that only a small fraction of the potential causes can be controlled in any given study. Measurement errors abound, and often even the directions of measurement biases will be subject to dispute. It is difficult enough to state nontrivial theories that are sufficiently precise to yield testable predictions. Specifying the nature of the conditions under which they may be expected to hold may be virtually impossible. Often, too, it is difficult to formulate theories that are overidentified, so that there will be an excess of empirical information that may be used to make consistency checks. For instance, if one allows for the possibility that X affects Y, and vice versa, one must either specify lag periods or introduce exogenous variables that affect one of the endogenous variables but not the other. Two social scientists may agree that reciprocal causation is involved but yet give different weights to the relative importance of the variables. They may also disagree on the specific identifying restrictions that would permit empirical estimation. Moreover, the necessary empirical data may be unavailable or too expensive or time consuming to collect.

There are therefore many reasons why our most important theories may be untestable as currently formulated, and not all of them stem from problems with the theories themselves. Yet, at least in sociology, it is difficult to reach any other conclusion than that this unsatisfactory state of affairs also results from the fact that many theories are vaguely worded, do not contain any predictive statements, and usually involve a sufficient number of ambiguously defined concepts and implicit assumptions that it is very easy to wiggle out of a set of embarrassing findings by invoking a variety of disclaimers. The situation is open to a diversity of interpretations and disagreements as to the adequacy of any given test or even a large number of negative findings.

What happens to an academic discipline when theories cannot be eliminated on empirical grounds? The field becomes cluttered with alternatives that must be evaluated in terms of some other criteria. One must pity not only the poor student who is asked to master such a range of theories but also the textbook writer attempting to present a reasonably complete picture of the field. Does one select "representative" theories, using some sort of judgmental criteria? The most recent theories? Those of the most prominent members of the profession? Those of the important historical figures? Or perhaps those of the dominant members of a particular department or one's mentors? How do students select among them? According to their ideological predispositions? The writers' styles? Simplicity? Or what they consider to be "politically wise" criteria, given the peculiarities of the department in which they are receiving their training?

Once the alternative theories become sufficiently numerous, it also becomes so difficult to attempt syntheses—or even summaries—that no one makes the effort. Each theory is studied superficially, except perhaps those of the dominant figures in the field. The motivation to locate definitive predictions that might provide crucial tests is also reduced, partly because there are so many alternative versions that few persons would be interested in such empirical results, given the inability to come out decisively in favor of one theory over another.

In sociology at least, the field of "theory" has been getting less prestigious and has even slipped so far as to have a bad name among many in the profession. This occurred despite the fact that we nearly all give lip service to the desirability of improving our theories and linking them more closely to empirical research. Token gestures appear in the form of passing references to theory in footnotes or by means of brief summaries of highly complex arguments. In many instances, however, the reader is left with the impression that such discussions are merely window dressing or are inserted so as to hint at an ideological position that may be fashionable at the time. For instance, a brief but positive reference to Karl Marx or one or two well-known phenomenologists, or perhaps a negative remark about Talcott Parsons or functionalism, serves to inform the reader that the writer is to the left of center within the discipline and, perhaps, among those who favor some "new" theoretical orientation to provide a backdrop for whatever empirical topic is being explored.

All this affects scholarly norms regarding the criteria by which theories are to be evaluated, the kinds of arguments one makes in critiquing or supporting a given theory, and how courses in theory are to be taught. If such criteria are primarily philosophical and if discussions remain at the level of highly abstract metatheoretical assertions, very little attention is likely to be given to what Merton (1968) termed "theories of the middle range." Nor will such abstract discussions seem at all relevant to theoretical issues that arise in specific subfields, as for example the study of complex organizations or social stratification. The utility of the theory simply is not one of the criteria by which it is judged. Problems of testability are set aside for others to study and are discussed in totally different contexts—in separate courses devoted to research methods.

Although I am unaware of any sociologist so rash as to study the matter empirically, one also suspects that there is self-selectivity with respect to the kinds of social scientists who elect to specialize in theory and in the departments that emphasize training in the area. If the field of theory is dominated in one department by traditionalists who stress the classics in the field, whereas in a second department it is dominated by phenomenologists or neo-Marxists, one would naturally expect that the students in the first department who elect to study theory may differ in important ways from those who do so in the second. Indeed, the field of theory is likely to become a battleground to be fought over by certain kinds of contestants, while most others simply retreat to more secure terrain where they can be left alone to do their own work. If such potential battlegrounds exist within a single academic department, conflict-reduction mechanisms are likely to result in a noncommunication strategy among the major parties. Students will be left to select whatever program of study is most compatible with their personal needs. Some may choose sides, a few may quietly "nibble" selectively according to their tastes, and many others will attempt to distance themselves as much as possible from anything to which the label "theory" has been attached.

NONPRODUCTIVE THEORETICAL DISPUTES

We must recognize that social scientists are human actors with individualistic goals that are not necessarily compatible with the long-term development of knowledge. Sometimes it is to one's benefit to keep a dying issue alive, to state a theory in such a fashion that it cannot possibly be rejected, or to set up straw men (they are rarely women) in order to oversimplify an argument taken to be in opposition to one's own viewpoint. Rather than seeing issues as challenges in which social scientists are working cooperatively as a team, we often perceive the situation as a zero-sum game in which it is either ourselves or our opponents who will discover the truth; if they are right, we must be wrong, or vice versa. There is also competition for readers, graduate students, journal space, and professional status.

In these respects all natural sciences, social sciences, and humanities share the same basic problems: how to advance knowledge, not only despite these human tendencies but by taking advantage of them. Where genuine knowledge contributions can be clearly recognized, the answer is obvious: Offer rewards to those who make the most important contributions. This device is difficult to employ, however, when such contributions are still in dispute, or when supposedly "new" discoveries or points of view cannot easily be distinguished from earlier ones because ideas are fuzzy, concepts have been redefined, or the scholarly literature is so vast that accurate accounting is virtually impossible.

First, we must recognize that many disputes stem from ideological differences—as, for instance, between those with leftist and Marxist orientations on the one hand and liberal to moderate ones on the other. Somewhat correlated with this left-right cleavage have been several others in sociology, at least. Blacks and then later other minorities led a challenge to what they perceived as "white" sociology, and this was soon followed by the radical feminist challenge to a male-dominated scholarly profession. As might be expected, these cleavages were also superimposed on generational differences, as well as an antiscience movement that stemmed from the Vietnam-era protest. The result was approximately 10 years of turmoil within the American Sociological Association, the various regional associations, many sociology departments, and—I am sure—the minds of individual sociologists.

Such challenges always have benefits, but they also leave their scars. On the positive side, we experienced considerable growth in the number of women in the profession, a more modest increase in minority representation, and far greater opportunities for both categories within professional societies and academic departments. Clearly, also, the growth and development of Marxist sociology afforded a healthy challenge to a number of lines of inquiry, particularly the fields of political sociology, stratification, race and ethnic relations, and criminology.

What these social movements also produced, however, was a greater legitimation for what I characterize as fruitless disputes and professional animosities. Even among colleagues, it became fashionable to use hate words: to call one another "racists," "male chauvinists," or "right-wing lackeys." It also became more legitimate to caricaturize whole studies by simple derogatory labels. Indeed, a few leading scholars were deliberately singled out as villains or heroes, the complexity of their views or the variety of their contributions ignored. In particular, Talcott Parsons and his more visible followers became the major symbol of establishment—and presumably conservative—sociology. "Functionalism" was singled out as the theoretical villain and "conflict theory" the hero.

In the field of race relations, white sociologists, as a category, were decried either as racists or as naive and therefore unqualified to teach courses about minorities. As might be expected, these acrimonious charges and countercharges produced personal conflicts within departments and within the professional associations. They also led to resignations from the American Sociological Association and the general withdrawal of those who could not or did not choose to take the punishment. Some white sociologists, for example, simply withdrew from the field of race relations. Others, who were not so timid and who in some instances deliberately encouraged the debates, were exceedingly willing to step in and also to fill the void produced in our scholarly publications. In the field of race relations this was a period of prolific writing and publisher interest in what we had to say. In my own judgment, however, it was not a period in which high-quality works were produced. Instead, it was a time of rhetoric and the devaluation of even the best works of the recent past.

Young scholars obviously have some vested interest in believing that previous work needs to be scrapped and that work of the establishment is seriously flawed or biased. Usually, of course, more established scholars have considerable power within their professions and so may counteract such tendencies. They control the journals, professional associations, and departmental decision-making processes, including the all-too-crucial decision to withhold or grant tenure. They also control the socialization of graduate students and their placement into initial jobs.

The 1960s were unusual in a number of ways. Opportunities for social scientists were rapidly expanding, there was still a shortage of graduate students. A relatively large number of junior-level faculty were in place within graduate departments, and their bargaining power for promotion for tenure was still considerable. There were also the near simultaneous social movements involving blacks, women, and antiwar protesters, all of which were highly relevant to sociology, political science, anthropology, and—to lesser degrees—the other social sciences. Furthermore, our professional norms encouraged debate, our guilt was enormous, and our intellectual standards were not clearly defined. We became a natural battleground for intellectual challenges to the system, as well as personal challenges to our sincerity as liberals and supports of the underdog. I believe we responded to the challenge fairly well and that the return to normalcy has been accompanied by positive gains. Our intellectual standards may have suffered as a result, however.

Ideologically based debates and cleavages are not the only sources of difficulty, however, as has been implied in the above discussion. Whenever a subject matter is highly complex and whenever arguments are diffuse, concepts unclear and strategies of attack by no means successful, it will be exceedingly difficult to distinguish between debates that will eventually bear fruit and those that will merely lead to endless bickering. Furthermore, it is often difficult for third parties to judge which debates are being prolonged primarily because of individual vested interests and which involve fundamentally differing assumptions about how to proceed or about the nature of social reality. In short, we are likely to be confused about how much attention to pay such debates. Perhaps there is something to them. Perhaps, however, we are merely wasting our time listening to them, apart from gaining vicarious satisfaction in seeing established figures shot down or senior professors calling each other names.

The tendency to oversimplify opponents' arguments and to set up straw men has existed for a long time within the social sciences. Sometimes this is a result of the way our theories are stated in the first place, making it exceedingly difficult to pin down underlying assumptions or to wade through pages of meandering discussions to get to the essential points being made. If an author tells us that, say, 10 specific assumptions are being made, the critic's job becomes much easier. At the same time, the honest critic will be called on to tell his or her own readers what the alternative assumptions are, in which case the dispute more readily becomes specific enough to result in a meaningful exchange. If the original author never says outright what he or she intended, however, and if a critic responds in like fashion, the temptation may be to try to persuade the innocent reader by colorful language, innuendo, or oversimplified labels.

Theory courses that include quick surveys of major theorists sometimes encourage this same kind of labeling. All I can remember about Gumplowitz, for example, is that he was a "conflict theorist." If one reads Pareto very quickly, as we were asked to do, one comes away with a few stock phrases such as "circulation of elites" or "lions and foxes." From here it is a simple step to the conclusion that lengthy books can be reduced to such slogans and that persons whose writings display a few commonalities can be lumped together into certain schools. Thus among students of social power we have "elitists" and "pluralists," and if one believes that societies are controlled by elites, the latter school of thought can be dismissed out of hand. Students may then be steered in the proper directions without having to bother themselves with such nonsense, with the derogatory labels sufficing to block out an awareness of serious scholarship with a very different orientation. Marxists were thus dismissed in sociology prior to the 1960s, whereas functionalists are similarly treated in some theory courses today. In one sociology department I visited, for example, undergraduate students had the impression that sociological theory and Marxism were one and the same thing.

One of the more indirect consequences of ideological or intellectual disputes of this nature is that they do not affect randomly all fields within a given discipline. This may result in a self-selective mechanism through which persons who wish to avoid them merely retreat from one area and enter another. As I have indicated, I rather suspect that many did this in the 1960s in the case of the field of race and ethnic relations. Radical feminists in sociology quite naturally entered the field of family and helped to create a new subfield of "sex roles' (now often referred to as "gender roles"). They also entered some subfields of stratification and political sociology. Marxists moved toward theory, as well as macrosociology and political sociology. Demography and human ecology, in contrast, continued to attract relatively apolitical, quantitative sociologists, whereas criminology-deviance became something of a battleground between traditional sociologists and Marxists. Once this kind of selective migration across fields occurs, it may take several decades to bring about a more reasonable balance, since fields become labeled as "female," "minority," "Marxist," "conservative," or "statistical." Research funding may thereby be influenced by the kinds of senior persons currently active in a given field, and this in turn helps to determine the kind of younger persons entering them.

Another disturbing possible outcome of such disputes is that they may encourage a form of intellectual timidity on the part of our best analytic minds, resulting in the overwhelming tendency to remain at the descriptive level and to refrain from either stating generalizations applicable to a diversity of situations or working with concepts involving a reasonably high level of abstraction. Given that many of us want our research findings to have practical applications and to be appreciated by the lay public, the task of drawing theoretical implications or seeing the larger picture is then left to others with the rather clear implication that they may do so at their own risk.

In sociology, at least, this abdication of responsibility for developing an intellectual core for the discipline has resulted in a proliferation of subfields, each of which is addressed to highly concrete social issues that are of considerable interest to a few but of very little concern to the overwhelming majority of the profession. We are retreating to the relative security of highly specialized niches, with the hope that somehow this will result in the accumulation of knowledge. It will, of course. But will the miscellaneous factual information we produce be of any real interest to social scientists two or three decades from now?

CAPITULATING TO OUR AUDIENCES

How we organize what we have to say and how we present our major ideas is obviously a function of the nature of our audience. Clearly, then, we are likely to communicate differently to our peers than to students or the lay public. Perhaps the majority of our journal articles are addressed to peer audiences and, since they are relatively short, can be packaged according to our own intellectual criteria. Not so with respect to most books we write, if only because publishers have an eye to profit and some larger audience. In contrast, research monographs can be focused more narrowly and autonomously organized. Thus the nature of our most important publication outlets, and their prevailing criteria for evaluating publications, will have important implications for the day-to-day writing behaviors of scholars in any given field.

For sociology and all the other social sciences, except in some areas of psychology and economics, our major outside audience consists of undergraduate students. Students, of course, invoke their own criteria for selecting courses and major fields, but these rarely involve the goal of advancing the intellectual quality of the discipline concerned. Many kinds of courses are considered valuable because they provide technical tools for work in other areas. Thus students take calculus courses because they will need to use this material in engineering courses. Elementary language courses have a similar pragmatic value, as do courses in statistics and research methods. Or students may wish to learn about the American political system, how our criminal justice system functions, or how to cope with marriage.

Presumably the majority of our other courses are taken for different reasons, but unless a student intends from the outset to become a professional social scientist, the motivation seldom centers on an eager desire to learn sociology, psychology, or even history. Students want to take courses that interest them, that deal with the kinds of relatively complex issues they are likely to encounter in adult life, or to prepare for a specific career for which they believe a background in that social science is relevant. This, in turn, implies that they are more likely to be attracted to courses referring to family life, race relations, crime and deliquency, urban politics, or Latin American economic development than to courses organized around more abstract social processes such as socialization, exchange and equity, and social differentiation and inequality. To the extent that academic deans look closely at enrollment figures and allocate faculty positions accordingly, social science departments can hardly afford to package their courses in unappealing ways, regardless of the strictly intellectual advantages of doing so.

A physics department might package its courses in a similar fashion: the physics of refrigerators, the physics of television sets, the physics of automobiles, and so forth. But because that discipline already has a strong theoretical core and a reasonably small set of *analytical* areas that focus on principles or laws of a more abstract nature, there is little temptation to organize the learning process around those concrete phenomena that might be most attractive to student interests. Instead, students are exposed to basic principles at very different *levels,* beginning with the most elementary. In terms of curriculum organization, this makes it far easier to build increasingly complex and technical learning sequences, with elementary courses serving as prerequisites for intermediate ones, and so on. Imagine, however, the difficulty of deciding on whether the course on refrigerators should precede or follow one on automobiles or television sets!

If one can learn about television sets only after first studying certain principles of electricity, it makes sense to discuss a diversity of applications, including a study of television sets, *after* the basic principles have been understood. Given four or five such applications, a student then can readily understand the relevance of the course on electricity. But how about the student interested in race and ethnic relations? How many different types of social processes would have to be studied in order to understand such highly complex social processes? It is hardly appropriate to put the student off until the senior year, especially if that student were also interested in a variety of other real-life processes. So we provide a course on race relations that contains a little bit of everything else—some social psychology of prejudice, an analysis of stratification, urban communities, the sociology of education, some indication of cultural diversity, a bit of economics, and some discussion of American history. The student then gets a repeat performance at the same elementary level when taking a course on the sociology of religion, the family, or criminology.

Textbooks must follow this lead, of course, with commercial publishers being closely attuned to matters of coverage and the necessity of not presuming too much background knowledge on the part of the average student. Rarely are students exposed to a second-level course in the same field, unless this be a seminar consisting of an eclectic body of readings. The result is an emphasis on breadth at the expense of depth. This would make sense in connection with a single survey course. When there is little or no increase in depth and where the same basic ideas are presented in elementary fashion in course after course, however, there is little sense of intellectual progress or a tough challenge to master really difficult material.

If our graduate-level courses are similarly constituted but simply involve additional readings and seminar presentations, the experience is repeated at a negligibly more advanced level. The result is a flat learning experience that is cumulative primarily in the sense of having to master a larger body of literature, rather than being one in which a genuinely deeper knowledge of the subject-matter is attained. Unless someone trained in the social sciences also has had the experience of being trained in a more technical subject, such as mathematics, he or she may not even recognize the qualitative differences involved. Rarely do good social science students find themselves intellectually over their heads, as frequently occurs within mathematics or physics.

Much of the same phenomenon occurs when we write books for lay audiences or research reports to be read by nonspecialists. These consumers of our knowledge are not interested in the theoretical development of sociology or political science as disciplines, but rather in some concrete social issue. They care little about general theories or the quality of the research supporting the propositions stemming from such theories. They want facts and straightforward interpretations, usually ones that square with both their common sense and vested interests.

Rather than take the trouble to explain the underlying theories—assuming they existed—the temptation is to provide mere caricatures of those theories we wish to dismiss and simplified versions of those we prefer. The result is a discussion that sounds like common sense, alongside a set of facts that have been stated in oversimplified fashion. More serious presentations are left for specialized articles or appendices that are seldom read. Nor is there much effort to relate the particular findings at hand to others in which the reader is unlikely to be interested. So attempts at generalization are not seriously made, except possibly in brief concluding chapters.

These and other defects in our intellectual cultures will not be easily corrected, if for no other reasons than inertia and a lack of consensus as to whether they constitute problems that need to be addressed. Nor can we assume that practices that are disadvantageous for the social sciences, as a whole, are also costly to a sufficient number of individual social scientists. It is important to keep in mind, however, that virtually all of the practices that have been discussed in the present chapter are under our own control. Unlike many of the methodological and theoretical complications discussed in the earlier chapters, those dealing with our own intellectual cultures *can* be addressed in a reasonably straightforward way. Whether or not we will choose to do so undoubtedly depends on the extent to which we are nearly all frustrated and disturbed by the present state of affairs. It will also depend on our ability to organize ourselves more effectively and to find ways of rewarding those who are willing to make the necessary effort. The final chapter concerns some of the positive steps that may be taken in this direction.

NOTE

1. See Blalock (1964: chap. 4) for further illustrations of this general problem of manipulating dependent variables.

CHAPTER
7

Some Positive Steps

What can be done, other than relying on the unfettered marketplace of ideas? How, if at all, can the knowledge cumulation process be speeded up and protected against the vagaries of intellectual fads and power struggles among rival schools of thought? Indeed, is it even desirable to attempt to modify the supposedly natural processes by which academic fields develop or atrophy? Can we count on some unseen hand to guide their development toward an idealized state of affairs, or is such a state totally unimaginable? What would it look like, and how could we assess the degree to which there are movements either toward or away from this idealized state? Most of us would probably agree that the physical and biological sciences are moving in such a direction, in many cases at an accelerating pace. But is it even meaningful to raise comparable questions in the case of the social sciences?

I believe that progress toward some objectives can be assessed, though it appears far easier to do so in some fields than in others. I also believe that there are important steps we can take to increase our rate of progress, though we may anticipate a number of tradeoffs and dilemmas that imply not only that our choices will be subject to debate but also that certain choices may entail costs as well as gains. If so, we must fully expect considerable debate as to their relative merits, as well as partings of the ways as some of us explore one path whereas others elect to follow another. What seems least satisfactory, in my view, is a laissez-faire stance that accepts the status quo as inevitable or as a natural consequence of the search for knowledge on the part of large numbers of individual social scientists. Such a position appears to ignore the possibility that, as social scientists, we may have evolved a subculture that contains many norms and beliefs that are dysfunctional to the knowledge development process.

Dissatisfaction with the current state of affairs seems absolutely essential for collective action and productive debate. If most of us accept an extreme lack of consensus and confusion over terminology as a necessary state of affairs, it may then be a rather small step toward believing it to be healthy or a sign of intellectual diversity and vigor. The more different positions taken by social scientists and the louder the debates, the better! A narrowing of perspectives, reaching agreements on basic concepts, and achieving a reasonable consensus on an intellectual core would, from such an orientation, be taken as threatening and an effort to impose orthodoxy. If students find our courses to be confusing, soft, easy, "Mickey Mouse" or repetitive, we may also turn a deaf ear or invoke the notion of academic freedom as an argument against serious reform efforts. Before long, we may also take the lack of an intellectual core as either an indicator of breadth or even a sign of maturity.

Such acceptance of the status quo is reinforced by the highly eclectic nature of our organizations. Most of us are members of academic departments or research institutes that are purposely developed to emphasize diversity for laudable purposes. For instance, we want to provide students with a set of course offerings that adequately represent an entire discipline, so that they may sample here and there according to their individual tastes. Even moderate- and large-sized departments normally contain only three or four specialists in each subfield, making sustained collaboration among departmental colleagues unlikely. Indeed, a university department that is successful in achieving overall excellence is highly unlikely to contain more than a handful of faculty who engage in genuine intellectual exchange with one another. Often their interactions take place primarily as a result of common ties with specific graduate students rather than through direct contact.

If other professional associations are at all similar to the American Sociological Association, they too will be incapable of sustained and systematic activities of a genuine intellectual nature. True, they may serve to facilitate exchanges through their journals, annual meetings, and sections. In some instances they may also constitute organizational vehicles through which small working groups of scholars may be encouraged, though usually on a short-term basis. But political considerations within such professional associations usually work to inhibit their taking forceful intellectual positions or encouraging one line of endeavor at the expense of another. The result is often wishy-washy endorsement of virtually every new activity, subgroup, or subfield that receives the support of a dozen or so members. Universalistic and anti-elitist values, though often functional in other important ways, carry over into the intellectual realm as well, so that almost every new idea is considered as valuable as any other. Challenges to orthodox positions are taken seriously regardless of their apparent merits, on the assumption that the professional association must remain scrupulously neutral and supportive of all new ideas. Thus almost everything flourishes, but to a limited degree. Members are given the opportunity to pick and choose from a maximum variety of options but also with maximum confusion. Undoubtedly, many may claim that this is just the kind of atmosphere a professional association should provide. It most certainly does not afford a sense of direction or encourage an emphasis on quality and depth, however.

Thus we must look elsewhere if genuine intellectual progress is to be fostered through any organized mechanism, or through the development of intellectual norms stressing the need for more effective communication and an efficient division of labor. At minimum, it is necessary to develop a set of priorities and reasonably general guidelines for achieving them and simultaneously to find ways to reward scholars for behaviors that seem most likely to increase the probability of moving in the right direction, while negatively sanctioning behaviors that inhibit others from adopting them. Finally, we must also achieve more effective organizational mechanisms that will help ensure *sustained* efforts that are relatively immune from faddish developments, political pressures, and the vagaries of the funding process. One cannot be optimistic that consensus on such objectives can be achieved rapidly, but until there is careful examination of alternative avenues, continual floundering is the most likely outcome.

DEVELOPING AN INTELLECTUAL CORE

The social sciences obviously differ with respect to the degree to which they focus on one or more core set of ideas. What I have in mind here is something similar to Kuhn's (1970) conception of a paradigm, although that notion, too, contains a number of ambiguities. The essential question to ask is this: To what degree do social scientists orient their empirical and theoretical work to a reasonably close-knit body of knowledge in such a way that the reader can understand the relevance of each particular work to such a larger body of knowledge? There is no suggestion, here, that there is necessarily a high degree of consensus on the truth of particular propositions or empirical facts relating to that body of knowledge, but whenever disagreements occur they are reasonably focused and the issues relatively clear-cut.

There are two essential ingredients here. First, there must be a reasonably general theoretical formulation stated without reference to time and place, but modifiable in terms of conditional statements that both limit its applicability and suggest how an even more general formulation may be developed by bringing in these conditional statements as additional variables into a more complex theoretical formulation that includes a number of important special cases.

The second ingredient consists of systematic efforts to check the theory with empirical information in such a way that modifications are brought about in a cumulative fashion. The theory (or a small number of alternatives) is *routinely* invoked to motivate nearly all empirical research in the field, and findings that are incompatible with particular formulations of the theory are treated as frustrating pieces of information that require immediate attention. Perhaps replication studies will produce results more compatible with the theory. Perhaps measurement artifacts will be discovered. The theory may be ambiguous in certain respects, so that crucial distinctions or conceptual clarifications will be needed. Where negative evidence continues to accumulate, a "paradigm revolution," in Kuhn's sense of that term, may occur, though the more usual case may simply involve a paradigm revision or clarification.

Clearly sociology is not characterized by these two ingredients, though many of us subscribe to them as an ideal to be approximated. Indeed, there may be a small number of subspecializations in which such focused research predominates, but the more usual state of affairs consists of a remote connection between theory and research. For instance, in the field of social stratification we have a large body of high-quality empirical literature on social mobility and status attainment. We also have an extensive theoretical literature involving Marxist and non-Marxist interpretations of inequality, the functions of stratifications and differentiation, the relationships among class, status, and power, and so forth.

These empirical and theoretical bodies of literature are rarely joined, however, and there is no sustained effort to bring the empirical evidence to bear on the general theoretical formulations or to use the latter to guide the former. For the most part the empirical literature has focused on describing the extent of occupational mobility in various countries, documenting status attainment processes and comparing those of white males, white females, and various minorities. This latter status attainment literature has proven to be cumulative in the sense that the empirical evidence has provided the opportunity to compare one group with another or to contrast several different time periods. The number of variables used in causal models of status attainment processes has also increased, and careful assessments of measurement errors have also been made. Thus there is considerable potential for the kind of cumulative paradigm-building effort we have been discussing. A meshing of this empirically based literature with other orientations and theoretical discussions in the stratification field seems a long way off, however.

What is most essential in this respect is the development of a set of norms and an intellectual climate that stress the importance of always asking this of any piece of empirical research: What *theoretical* questions or issues are being addressed, and how are these questions linked to some larger body of theoretical knowledge in this field? How can a particular study—which necessarily deals with some concrete set of data referring to a specific time and place— be assessed in terms of the light it sheds on theoretical questions that have been formulated in more general terms? Does the author of the empirical report make this relationship to the theory clear, or must the reader extract theoretical implications with considerable effort? Do we collectively produce a body of literature in which there is considerable concern with and frustration about theoretical puzzles arising from incompatible empirical findings? Or is the flavor more one of miscellaneous empirical findings that are interesting only because they happen to involve timely social issues or matters that are directly relevant to a specialized readership?

The picture is uneven in sociology and, I presume, in other social sciences as well. As implied, the empirical literature on status attainment processes is, in my opinion, one of our bright spots, since it offers at least the potential of being integrated with theoretical issues of interest to other scholars in the field of social stratification. We have similar though limited bright spots elsewhere in the discipline. One can hope that these might serve as exemplars for the rest of us, but we also hear many comments that suggest the contrary. I have heard the statement, "We don't want to hire anyone who has 'merely' added a variable or two to the Blau-Duncan (1962) model of status attainment." Others have implied that the status attainment literature has become boring.

Criticisms of any approach will be inevitable and are needed. What I perceive as lacking in the criticisms of the type referred to, however, is any stress on the need for a cumulative effort that seeks to *link* two or more bodies of literature in a constructive fashion. If one were to ask a series of questions oriented to expanding and generalizing a body of literature, as exemplified by research in the status attainment field, and if a substantial group of specialists were then to begin addressing such questions, several minicore theoretical areas might be joined. Perhaps the issues raised would then be seen as constituting a frustrating and challenging set of puzzles to be solved, so that more general theoretical concerns helped to dictate the choice of empirical problems to be investigated.

Putting the theoretical horse before the empirical cart makes sense, however, only when one makes sure that the two are hitched in a secure though flexible manner. This implies that the important concepts in the theory have been carefully defined, with sufficient attention paid to the implied measurement operations that reasonably unidimensional variables and comparable procedures can be used across a diversity of settings and periods. Too often—in sociology at least—lengthy theoretical discussions and debates take place without any reference to the quality of the empirical evidence or even the kinds of data that would have to be collected to provide suitable measures of the important variables and tests of the theoretical propositions at issue.

Indeed, the theoretical discussions may not contain any explicit propositions at all. Readers may be only dimly aware of their implicit existence, buried somewhere among colorful metaphors, anecdotal evidence, polemical statements, and summaries of historical materials. Most disturbing of all, in my view, is the fact that many social scientists apparently accept uncritically the view that theory must inevitably take this form and that the burden rests with empirical researchers to operationalize the concepts in the theory in whatever ways they can, without much help from persons who refer to themselves as theorists.

The specific problems faced by each social science discipline are somewhat unique, given the strengths and weaknesses of theory development in each area, measurement and data collection problems peculiar to the field, and the scholarly norms and expectations that have evolved. Also, each discipline tends to have its own set of blinders or biases that inhibit its maturation. The suggestions that follow are therefore general and based primarily on my own experiences in sociology and, to a lesser extent, my nodding acquaintance with developments in political science and psychology. Each will be mentioned only briefly in this concluding chapter, as each has been alluded to in previous discussions.

The first and perhaps most important suggestion is that the major orienting principles in each discipline be organized in terms of social processes that are sufficiently general in nature that they may be described and analyzed in terms of propositions that are not tied to particular time- and space-bound entities such as specific nations, racial or ethnic groups, categories of individuals, and named occupations or organized groups. For instance, psychologists tend to organize their thinking around processes such as learning, cognition, perception, attribution, motivation, and development. Sociologists, social psychologists, and economists study exchange and equity processes; race relations specialists study segregation and approach-withdrawal processes and discrimination, and so forth. As noted in Chapter 5, each of these pure processes may be analyzed in formal terms, in much the same way that Simmel (1950) advocated. Since they will generally occur jointly in real-world settings, however, it then becomes necessary to specify how each simple process becomes modified when it is combined with another—as, for example, when exchange processes are complicated by power differentials or when approach and withdrawal processes also involve exchange relationships.

An investigator studying any reasonably complex phenomenon will, as I have suggested, have a difficult time relating this phenomenon to any theory about a *single* process unless there are substantial hints as to how this theory will have to be modified whenever other processes are simultaneously involved. Without such elaborations on the simpler theories, an empirically minded investigator may be faced with an all-or-nothing situation in which the theory either works in highly simplified settings or breaks down in more complex ones. Not wanting to constrain the observations in a grossly oversimplified fashion, the only way out may appear to be that of resorting to detailed descriptions plus passing reference to the inadequacies of existing theories. Thus the hiatus between theory and research is preserved, with empirical evidence for or against the theory being primarily anecdotal and noncomparable from one study to the next. The implication is that theoretical accounts of social processes need to be made more complex if we are ever to transcend the kind of evidence available in simple laboratory settings.

My second suggestion is that far greater attention needs to be given to problems of measurement and conceptualization, particularly in instances where measurement is highly indirect and where lack of measurement comparability is likely to inhibit the testing of general theories in a diversity of settings. This suggestion is also not a new one; I have elaborated on it in considerable detail elsewhere (Blalock, 1982). The fact is, however, that at least in sociology and political science only a minor portion of our total energy has been devoted to measurement and conceptualization problems. This is precisely what one would expect whenever a discipline is attempting to cover too much territory and is therefore spread too thin over such a wide range of practical topics that each receives only superficial coverage. Without a core set of questions and a reasonably small number of important variables to examine, we cannot expect members of a discipline to be encouraged or rewarded for in-depth and rather mundane attention to the subtleties of measurement, to replication and reliability studies, or even to multiple pretests of one's data collection instruments. One must move on to the next study so as to sustain one's funding or to gain tenure or promotion.

The codification of reasonably general theoretical propositions is of crucial importance in providing applied researchers with ready access to whatever theories may be relevant to the topic at hand. Obviously, useful codification requires a reasonably uniform vocabulary of terms. It also would be helpful to provide discussion of alternative measures or indicators that have been used in other research, along with the assumptions that are most commonly made in linking them to the theoretical constructs they are intended to measure. Citations of methodological discussions, reliability checks, and empirical findings would likewise be useful to applied researchers. Unless the number of concepts or variables taken to be basic is kept to a reasonable number, however, the sheer effort in producing initial codification schemes would be overwhelming, to say nothing of the paucity of rewards for such an undertaking. Until serious efforts are made to list and scrutinize important variables, we may anticipate that the codification of propositions and research findings will remain a nearly impossible dream. So will be the objective of making theoretical ideas directly relevant to applied research.

Third, it seems important to stress norms that encourage investigators to work back and forth, self-consciously, between the general and the specific. If, for example, everyone doing a case study were to conclude with a careful analysis containing specific theoretical propositions and indicating precisely how that case study might be relevant to others in similar settings, we would be in a much better position to build each study on top of preceding ones. If one is studying power relationships within family settings, why not state a series of propositions that would apply to other small groups and then attempt to specify, in general terms, the variables that distinguish families from these other groups? Or if one is studying interaction processes within a particular bureaucratic setting, why not select two or three other different settings for comparison purposes? I am not implying that authors never actually look beyond their own data sets, but we do not do so on a *regular* basis, as a precondition for the acceptance of our research by the community of scholars.

Nor do we self-consciously or routinely cross levels of analysis by comparing individual-level data with aggregated data. Contextual effect or cross-level analyses are still rare in most fields of sociology and political science and—as far as I know—are virtually absent in psychology and anthropology. The methodological problems in such cross-level analyses are subtle but tractable, and, as already noted, there is now an extensive body of literature on the subject. Two major difficulties stand in our way. The first is highly practical and formidable: Data collection costs are substantial, and considerable coordination of research efforts would be needed over a prolonged period. The second is more normative and a function of the manner in which disciplines and subfields have been defined. It may rarely occur to a macrosociologist to examine individual-level data, nor to a social psychologist to insert community-level data as explanatory variables in an individual-level analysis. What we need is a set of expectations that reward and routinely encourage cross-level analyses whenever data at more than one level can be obtained with reasonable effort.

Fourth, we need to find ways to encourage systematic searches for data gaps, especially when these gaps have resulted from the overuse of specific data collection techniques and the underuse of others. If some substantive fields rely heavily on survey research, whereas others rely primarily on laboratory experiments, what gaps does this produce? Do certain theoretical fads get linked with particular data-gathering techniques that are also selective with respect to the kinds of variables that can be measured? Are these techniques such that certain kinds of variables are measured more accurately than others, so that there will be differential measurement errors? What are the implications of omitting all but contemporary values of certain variables in terms of the degree of support that may be expected for one kind of theory as compared with another? Suppose we discover that macro-oriented scholars, say in the field of race relations, totally ignore the findings of social psychologists and vice versa? Could this be a function of the kinds of data each has been examining? Does it result from the existence of major data gaps that, if closed, might bring the two points of view closer together? Who will do the searching for such data gaps and with what professional encourgement? If we make no conscious effort to fill the gaps, will they be filled at a later time as a result of some "natural" process? What do we do in the meantime, given the gaps? I am afraid we often simply ignore them and hope they will disappear.

Fifth, we need to find more effective ways of encouraging and rewarding certain kinds of routine activities that contribute to the cumulation of knowledge, even though they are not particularly exciting or intellectually challenging to persons who are actively engaged in research. In particular, we often complain that research findings are seldom replicated or at least that replication research does not find its way into print and is not professionally rewarded. Of course not all pieces of research need to be replicated, and so ways must be found to encourage replications in instances where the research findings were especially relevant to important theories or where replications would serve to provide knowledge as to the conditions under which a given set of results will or will not obtain. The key notion here is that of *guided* replication research. Once more we see that the crucial ingredient is a sense of what is and is not important in terms of theory development.

What kinds of persons should be encouraged to engage in replication research? One obvious set of candidates comprises students, since replication research is often useful as a learning experience and can also be exciting at that stage. To the extent that data collection is expensive and time-consuming, however, student research may not be suited for this purpose, though in instances where preexisting data sets are available for use, students may at least replicate the analysis portions of the research. Many social science faculty are located in undergraduate institutions that place heavy instruction burdens on them while at the same time expecting at least a modest research effort as well as tangible products in the form of publications. Especially in instances where these faculty can make use of student assistants who will benefit from the research experience, replication research seems an ideal activity. It must be rewarded, however, both by the profession as a whole and by the employing institution. Insofar as these institutions require tangible publications, the existence of specialized journals encouraging replication research would be advantageous. More useful, perhaps, would be sections within existing journals devoted to brief discussions of results of replication studies. What we need, it seems, is both recognition of the importance of replication research and a set of procedures enabling us to facilitate the reporting of such research and rewarding those who do this important work.

There is also a need to encourage and reward certain other kinds of activities such as archiving and computer consulting. At many universities the latter role is often performed by graduate students or programmers who lack doctorates, and who are therefore vulnerable to budgetary cutbacks or who are locked into dead-end positions that are not adequately rewarded in terms of the usual academic criteria. Although data archiving is a more specialized activity that is carried out at a smaller number of institutions, the same kind of situation prevails—namely, insufficient academic rewards and job insecurity in times of budget cutbacks. We do not seem to be organized in such a way as to institutionalize such roles except on a local and often ad hoc basis. It seems unlikely that we will act to change this situation unless and until these important roles become recognized as crucial to the accumulation of knowledge.

Finally, we need to give careful attention to the implications of our graduate training programs for the kinds of orientations we encourage among future social scientists. Do we encourage them to think in terms of *programs* of research that extend over five years to a decade? Or do we reward them primarily for doing isolated pieces of research? Are they expected to think through carefully the potential contributions to theory that a given research project is likely to make, or do they merely jump into some ongoing research being conducted by their mentors? Are they ever given the experience of "fighting through" a difficult intellectual task, from an initial groping period to a more routinized set of activities, to thinking through its implications for knowledge accumulation in other subfields? I am continually dismayed by the high proportion of job candidates in sociology, all coming from top departments, who appear never to have thought much about the connection between their own research and that in other subfields, or even the interconnections among the research projects with which they have been associated.

Are we turning out research technicians on the one hand and broad-gauged theorists on the other, without much blending of the two? How do we provide students with increasingly necessary technical tools without sacrificing general theoretical training? Undoubtedly the answers to these questions will differ by discipline and across departments, but I believe they need to be answered or at least seriously addressed. This is especially crucial during the coming decade when the absolute numbers of academic positions in the social sciences will be decreasing, but where we may also hope to compensate for these trends by producing increasing numbers of social scientists with highly applied orientations.

DEVELOPING AND CLARIFYING
SCHOLARLY NORMS

I believe that we need to develop and implement a set of scholarly norms designed to improve communication both within and across the several social sciences. Basically, we must give priority to norms that stress the importance of intellectual honesty and integrity, completeness and open-mindedness. To the degree that the social scientist is careful to state an argument as clearly as possible, defining key concepts explicitly, providing the reader with a list of assumptions, and then going on to point to inadequacies and needs for specific kinds of follow-up studies, the role of the critic is being both anticipated and facilitated. Furthermore, reasonable critics are far less likely to become angered by qualified assertions than dogmatic ones, nor will they be as frustrated as would be the case if implicit assumptions have been buried in wordy discussions. I presume they would be more inclined to take the qualified findings or theoretical assertions as tentative building blocks rather than as foundations that need to be destroyed before taking a totally new starting point.

Norms for critics, in turn, need to stress the desirability of careful study and reporting of the original arguments or study rather than a superficial caricature of its thesis. This would include a genuine effort to look for strong points as well as weak ones, and to bring these to the attention of the reader by providing a rather complete account of the thesis or study that is being criticized. Sarcasm, simplistic slogans, guilt by association, and failures to mention important qualifications made in the work can all be negatively sanctioned by a critic's readers if they are so inclined and if their socialization into the profession has stressed that such tactics are not part and parcel of scholarly criticism. Of course, we all recognize the temptation to pass off certain whole lines of research in this fashion and so also tend to reward others for poking fun at such research. Furthermore, in some instances we really believe these lines of inquiry to be fruitless, pretentious, or at least marginal to the development of the discipline. Therefore it is difficult to apply such norms of disapproval in a universalistic fashion, and we may secretly applaud intellectual kindred spirits who do the dirty work for us. Unfortunately, however, the tactic can be applied by one's "opponents" as well, and escalation is likely to result.

The norms we try to instill in connection with statistical analyses have, I think, considerable applicability to this kind of issue. There is something about the way we go about analyzing data with a multiple regression equation that suggests the kinds of norms to which I refer. We try to make our assumptions explicit. We recognize that there are many omitted causes, and, ideally, we try to name these and suggest how they may be measured in future studies. We must include them in our error terms, however, and so we also make explicit assumptions about these error terms. A reader is thereby sensitized to these assumptions and challenged to include more variables. We also call the reader's attention to possible measurement errors and to other complications such as nonlinearity, nonadditive joint effects, or high intercorrelations among independent variables. We allow for such complications in our analyses and tell the reader how much additional variance they explain. Where independent variables are intercorrelated, we try to construct causal models accounting for these correlations, and we admit to overlapping "explained variance" that cannot be partitioned among independent variables without rather strong assumptions.

Most important, when some variables don't work, we provide the evidence along with numerical measures of the degree of explanatory power of each variable. We also try to warn readers about various artifacts due to the research design, peculiarities of our samples, restrictions placed on data collection, and—of course—sampling errors. I believe that most of us who teach statistics also stress to our students the need for intellectual honesty at all stages in the process. What also should come through is the ideal of value-neutrality with respect to selection of independent variables, as well as the presentation of research findings.

I can fully understand the skepticism toward quantitative research among those who honestly believe that the most important explanatory variables in which they are interested cannot be easily measured or that considerably more exploratory research would be required before one should attempt studies that include only a small fraction of these variables. Specific quantitative studies may be criticized on any number of grounds and theoretical formulations based on such studies will necessarily be incomplete. It seems to me, however, that the *norms* suggested by statistical research are reasonable ones for scientific analysis in general. Furthermore, a careful study of the literature on multivariate analysis leads one to a specific set of guidelines that are stated well enough that they may be followed, replicated, and evaluated by third parties. Such guidelines, of course, do not carry over directly to the evaluation of theoretical arguments or nonquantitative research. But it may not be too naive to believe that students who have been encouraged to incorporate norms of objectivity with respect to statistical data analyses will also attempt to carry over similar orientations to less quantitative research as well. I certainly have heard no persuasive arguments to the contrary.

We also need to modify a number of stylistic norms that derive partly from the humanistic and literary orientations of many of our profession but that seem incompatible with those of scientific reporting, though they may be highly appropriate in more journalistic works directed toward lay audiences. To make a book interesting, it may be important to use colorful words, to avoid summaries and repetitive terminology, to minimize the use of tables and equations or to replace them with more readily understood figures and charts, and to relegate technical arguments to appendices or delete them altogether. It is also more colorful to attack one's opponents, to make "cute" derogatory remarks that encourage the reader to dismiss their arguments without much thought, and to exaggerate the significance of one's own findings. Book reviews designed to "dethrone" a particular theoretical point of view, rather than to summarize the basic content, are in much the same tradition, and they undoubtedly make for more enjoyable reading.

To the degree that social scientists read each other's works with the expectation of being entertained, excited by entirely new and provocative arguments, or provided with interesting reading material for undergraduate courses, we encourage much the same type of writing style. Scientific expository writing is indeed dull. The excitement—if one must use that term—comes from the ideas themselves and from the mastery of a complex set of information rather than the writing style itself. Of course, we might like to have it both ways. A few writers do have the ability to communicate effectively without relying on stylistic gimmicks or the intellectual counterpart to social gossip. To the degree that we fail, however, to stress the importance of clear, concise, and simple expository writing, and a corresponding emphasis on the consistent use of terminology, explicit definitions, periodic summary statements, and explicit lists of variables and propositions, we are also failing to emphasize the very great need for effective communication *within* our own academic disciplines.

ORGANIZING THE EFFORT

How can we proceed if academic departments and professional associations are not the likely vehicles for promoting effective intellectual exchanges and coordination among social scientists interested in furthering the development of analytic areas? Clearly it is not feasible to attempt to bring such persons together, on a long-term basis, within single departments or even our large universities. Nor is it likely that sufficient sustained funding will become available to establish research institutes, unless these have a highly applied focus or are opportunistically eclectic in terms of the research grants and contracts they seek in order to support a professional staff. Several other models come to mind, however, and no doubt more can be conceived if a sufficient number of interested scholars can be induced to take the matter seriously.

One model is the "think tank," such as the Center for Advanced Studies at Palo Alto, that attracts scholars who wish to spend their sabbatical leaves in an atmosphere where informal communication with other social scientists is facilitated. Let us imagine that five to eight analytical areas could be designated and reasonably clearly delimited. If it were known that topics would be rotated on a regular schedule, sabbaticals could be planned and individuals invited so that the participants in any given year were all interested in a given analytic area, though their academic disciplines and applied interests might be diverse. One would expect that, within a short period, the value of such opportunities would become so obvious that scholars would begin to plan years in advance to take their leaves to be able to participate in such programs. Those who could not arrange a prolonged leave would presumably be able to take part in more delimited exchanges—for example, two-week workshops held at convenient intervals during the year.

Another model is the summer training program that combines course offerings and retooling opportunities for both students and established scholars with more intensive seminars for a smaller number of scholars whose work is closer to the "cutting edge" of the analytic topics. These latter scholars may, of course, constitute the instructors for the training program. Advanced graduate students specializing in the area may be involved in the seminars and also serve as assistants in the training program. The excellent summer program in quantitative research and applied statistics conducted by the Inter-University Consortium for Political and Social Research (ICPSR) is an example of such a program that has had a substantial impact over the course of nearly two decades.

A somewhat less ambitious model, and one that seems especially appropriate as a "bootstrap" type of operation, is that of the brief working session that may take place over a period of several days, perhaps in conjunction with the annual meetings of the major professional associations. If such working sessions are one-shot affairs that involve no follow-ups or assignments, we anticipate that they will have a negligible impact. If conducted annually, however, and with specific objectives in mind, such as the publication of a volume of research papers or the establishment of a specialized journal, even this relatively low degree of coordination may serve as a useful catalyst for other activities of a more sustained nature. Within sociology, for example, a relatively small group of scholars interested in social networks has effectively used this simple type of organizational device to create a reasonably coherent network of network specialists as well as a journal devoted to the topic.

Organizing without a reasonably delimited focus obviously makes little sense. Finding *analytic* criteria that can serve as really fruitful bases for coordination and intellectual exchange is undoubtedly more difficult than using applied foci. Funding for applied research institutes (in noncontroversial areas) is certainly easier to obtain, especially during the current period of contraction of resources and lowered priorities for basic research. Therefore unless we begin on a small scale and with modest objectives, we are likely to be disappointed with the results.

What if we do not make the effort at all, however? Will we be satisfied to drift or to let forces over which we have little or no control determine the shape of the social sciences in the twenty-first century? Can we predict what that shape will be? We all have our hunches or hypotheses. My own are not sufficiently optimistic to justify such a laissez-faire stance. I have therefore tried to argue in this book that we need to give considerable thought to the important questions about where we are heading, where the impediments to genuine progress lie, and what steps may be taken in whatever directions we, ourselves, choose to move.

References

179

Becker, H. Through Values to Social Interpretation: Essays in Social Contexts, Actions, Types and Prospects. Durham, NC: Duke University Press, 1950.

Berger, P. Invitation to Sociology: A Humanistic Perspective. Harmondsworth, England: Penguin, 1971.

Blalock, H. M. Causal Inferences in Nonexperimental Research. Chapel Hill: University of North Carolina Press, 1964.

Blalock, H. M. "Causal inferences in natural experiments: some complications in matching designs." Sociometry 30: 300-315, 1967.

Blalock, H. M. Conceptualization and Measurement in the Social Sciences. Beverly Hills, CA: Sage, 1982.

Blalock, H. M. and P. H. Wilken. Intergroup Processes: A Micro-Macro Perspective. New York: Free Press, 1979.

Blau, P. M. Exchange and Power in Social Life. New York: John Wiley, 1964.

Blau, P. M. and O. D. Duncan. The American Occupational Structure. New York: John Wiley, 1967.

Boyd, L. H. and G. R. Iversen. Contextual Analysis: Concepts and Statistical Techniques. Belmont, CA: Wadsworth, 1979.

Box, G.E.P. and G. M. Jenkins. Time Series Analysis: Forecasting and Control. San Francisco: Holden-Day, 1976.

Campbell, D. T. and D. W. Fiske. "Convergent and discriminant validation by the multitrait-multimethod matrix." Psychological Bulletin 56: 81-105, 1959.

Christ, C. Econometric Models and Methods. New York: John Wiley, 1966.

Cicourel, A. V. Method and Measurement in Sociology. New York: Free Press, 1964.

Coombs, C. H. "Theory and methods of social measurement," in L. Festinger and D. Katz (eds.) Research Methods in the Behavioral Sciences. New York: Dryden Press, 1953.

Coombs, C. H. A Theory of Data. New York: John Wiley, 1964.

Cox, O. C. Caste, Class, and Race. New York: Doubleday, 1948.

Denzin, N. K. The Research Act: A Theoretical Introduction to Sociological Methods. Chicago: Aldine, 1970.

Dollard, J. Caste and Class in a Southern Town. New Haven: Yale University Press, 1937.

Drake, St. C. and H. R. Cayton. Black Metropolis. New York: Harcourt Brace Jovanovich, 1945.

Duncan, O. D. "Path analysis: sociological examples." American Journal of Sociology 72: 1-16, 1966.

Duncan, O. D. "Some linear models for two-wave, two-variable panel analysis." Psychological Bulletin 72: 177-182, 1969.

Duncan, O. D. Introduction to Structural Equation Models. New York: Academic Press, 1975.

Duncan, O. D. and B. D. Duncan. "A methodological analysis of segregation indexes." American Sociological Review 20: 210-217, 1955.

Emerson, R. M. "Exchange theory, part II: exchange relations and network structures," in J. Berger, M. Zelditch, and B. Anderson (eds.) Sociological Theories in Progress. Boston: Houghton-Mifflin, 1972.

Etzioni, A. "Social analysis as a sociological vocation." American Journal of Sociology 70: 613-622, 1965.

Farkas, G. "Specification, residuals, and contextual effects." Sociological Methods & Research 2: 333-363, 1974.

Goldberger, A. S. Econometric Theory. New York: John Wiley, 1964.

Greenberg, D. F. and R. C. Kessler. "Equilibrium and identification in linear panel models." Sociological Methods & Research 10: 435-451, 1982.

Hammond, J. L. "Two sources of error in ecological correlations." American Sociological Review 38: 764-777, 1973.

Hannan, M. T. Aggregation and Disaggregation in Sociology. Lexington, MA: D. C. Heath, 1971.

Hannan, M. T. and L. Burstein. "Estimation from grouped observations." American Sociological Review 39: 374-392, 1974.

Hanushek, E. A. and J. E. Jackson. Statistical Methods for Social Scientists. New York: Academic Press, 1977.

Hauser, R. M. "Contextual analysis revisited." Sociological Methods & Research 2: 365-375, 1974.

Heise, D. R. "Causal inference from panel data," in E. F. Borgatta and G. W. Bohrnstedt (eds.) Sociological Methodology 1970. San Francisco: Jossey-Bass, 1970.

Heise, D. R. Causal Analysis. New York: Wiley-Interscience, 1975.

Johnston, J. Econometric Methods. New York: McGraw-Hill, 1972.

Jöreskog, K. G. and D. Sörbom. LISREL V: Analysis of Linear Structural Equation Systems by Maximum Likelihood and Least Squares Methods. Chicago: National Educational Resources, 1981.

Krantz, D. H., R. D. Luce, P. Suppes, and A. Tversky. Foundations of Measurement, Vol. 1. New York: Academic Press, 1971.

Kuhn, T. S. The Structure of Scientific Revolutions. Chicago: University of Chicago Press, 1970.

Land, K. C. "Principles of path analysis," in E. F. Borgatta (ed.) Sociological Methodology 1969. San Francisco: Jossey-Bass, 1969.

Langbein, L. I. and A. J. Lichtman. Ecological Inference. Beverly Hills, CA: Sage, 1978.

Loftin, C. "Galton's problem as spatial autocorrelation: comments on Ember's empirical test." Ethnology 11: 425-435, 1972.

Long, J. S. "Estimation and hypothesis testing in linear models containing measurement error: a review of Jöreskog's model for the analysis of covariance structures." Sociological Methods & Research 5: 157-207, 1976.

Lundberg, G. A. Can Science Save Us? New York: Longmans, Green, 1961.

Lynd, R. S. Knowledge for What? The Place of Social Science in American Culture. Princeton: Princeton University Press, 1939.

Merton, R. K. Social Theory and Social Structure. New York: Free Press, 1968.

Merton, R. K. "The precarious foundations of detachment in sociology: observations on Bendix's sociology and ideology," in E. A. Tiryakian (ed.) The Phenomenon of Sociology. New York: Appleton-Century-Crofts, 1971.

Mills, C. W. The Sociological Imagination. New York: Oxford University Press, 1959.

Myrdal, G. An American Dilemma. New York: Harper & Row, 1944.

Myrdal, G. Objectivity in Social Research. London: Duckworth, 1970.

Namboodiri, N. K., L. F. Carter, and H. M. Blalock. Applied Multivariate Analysis and Experimental Designs. New York: McGraw-Hill, 1975.

Naroll, R. "Galton's problem," in R. Naroll and R. Cohen (eds.) A Handbook of Method in Cultural Anthropology. Garden City, NY: Natural History Press, 1970.

Nelson, C. R. Applied Time Series Analysis for Managerial Forecasting. San Francisco: Holden-Day, 1973.

Robinson, W. S. "The logical structure of analytic induction." American Sociological Review 16: 812-818, 1951.

Schoombee, G. F. "The act of value-judgement: the sociologist's dilemma." South African Journal of Sociology 14: 26-34, 1983.

Shingles, R. D. "Causal inference in cross-lagged panel analysis." Political Methodology 3: 95-133, 1976.

Simmel, G. Sociology of Georg Simmel (K. H. Wolff, trans.). New York: Free Press, 1950.

Taeuber, K. E. and A. F. Taeuber. Negroes in Cities. Chicago: Aldine, 1965.

Tuma, N. B., M. T. Hannan, and L. Groeneveld. "Dynamic analysis of event histories." American Journal of Sociology 84: 820-854, 1979.

Weber, M. The Methodology of the Social Sciences. New York: Free Press, 1949.

Znaniecki, F. The Social Role of the Man of Knowledge. New York: Harper & Row, 1968.

About the Author

Hubert M. Blalock, Jr. received his Ph.D. in sociology from the University of North Carolina in 1954. He is a past President of the American Sociological Association and Fellow of the American Statistical Association, and has been elected to the National Academy of Sciences and the American Academy of Arts and Sciences. The author of a number of books, including *Social Statistics, Theory Construction,* (with Paul H. Wilken) *Intergroup Processes: A Micro-Macro Approach,* and *Conceptualization and Measurement in the Social Sciences,* he is currently Professor of Sociology and Adjunct Professor of Political Science at the University of Washington, having previously taught at the University of Michigan, Yale University, and the University of North Carolina.